Schedules for Clinical Assessment in Neuropsychiatry

Version 2

MANUAL

Present State Examination
Item Group Checklist
Clinical History Schedule

World Health Organization
Division of Mental Health
Geneva

Contents of SCAN Manual

CLINICAL HISTORY SCHEDULE (CHS)

Section 27

Acknowledgments

The SCAN system (Schedules for Clinical Assessment in Neuropsychiatry) has been developed in the framework of the World Health Organization (WHO) and National Institutes of Health (NIH) *Joint Project on Diagnosis and Classification of Mental Disorders, Alcohol and Drug Related Problems* (Principal Investigator, *N Sartorius*, WHO). The development of SCAN was funded by WHO, NIH, and Institutes employing collaborators who took part in the project.

SCAN had its origins in an existing instrument, the Present State Examination (PSE). A brief introduction to the earlier history of the PSE, to the structure of SCAN, and to the development of other instruments in the Joint Project, will be found below.

SCAN was developed in three phases:

1982 - 1990 Development and field trials of a pilot version
1990 - 1992 Revision for change in ICD-10 DCR. Production of a printed pre-publication SCAN Version 1, WHO, Geneva.
1992 - 1994 Updating to incorporate the final revision of ICD-10 DCR and DSM-IV items and algorithms. Publication of SCAN Version 2.

It is impossible to acknowledge the work of everyone who has participated during the past 10 years. No such enterprise can succeed without the advice, support and collaboration of many researchers, centers and agencies. Some of them are listed below.

SCAN Version 1 (1982-1992)

WHO: *N Sartorius, A Jablensky, TB Üstün, M Grant*
NIH: *J Blaine (NIDA), J Burke (NIMH), B Grant (NIAAA), R Hirschfeld (NIMH), D Regier (NIMH), L Towle (NIAAA)*

Task Force on Diagnostic Instruments: *JK Wing* (chair), *M von Cranach, C Pull, L Robins, H-U Wittchen,* investigators from Field Trial Centers, and staff of WHO and NIH.

Field Trial Centers, Principal Investigators: (See end of SCAN Manual volume for list of participants and addresses.)

Ankara, Turkey: *A Gögüs*
London, UK: *P Bebbington*
Athens, Greece: *V Mavreas*
Lübeck, Germany: *H Freyberger*
Bangalore, India: *M Isaac*
Mannheim, Germany: *K Maurer*

Beijing, China:	*Shu Liang*
Nottingham, UK:	*G Harrison*
Canberra, Australia:	*B Hughson*
Santander, Spain:	*JL Vazquez-Barquero*
Farmington, USA:	*J Escobar, T Babor*
Sao Paolo, Brazil:	*L Andrade*
Geneva, Switzerland:	*L Barrelet*
Sofia, Bulgaria:	*V Nikolov*
Groningen, The Netherlands:	*W van den Brink*
Sydney, Australia:	*G Andrews*
Leicester, UK:	*T Brugha*

Contributions to the design and trial of particular sections of SCAN:

Somatoform:	*P Morosini, TB Üstün*
Obsessional:	*G Andrews*
Alcohol and drug:	*T Babor, A Bertelsen, A Gögüs, B Grant, V Nikolov,J Strang, T Tomov, JL Vazquez-Barquero*
Cognitive:	*V Mavreas, JL Vazquez-Barquero, M Roth*
Social impairment:	*LG Wing*
Glossary:	*T Babor, P Bebbington, R Campbell, TB Üstün*
General editing and improvement:	*P Bebbington, A Bertelsen, T Brugha, J Cooper, J Escobar, A Farmer, V Gentil, R Giel, AS Henderson, A Jablensky, TB Üstün*
Computer design and programs:	*G Der, N Contractor, S Gauthier, E Glover*
Computer assisted interview:	*G Glover*
Consultants:	*H Pfister, D Rae*
Print and graphic design:	*Design Locker (M Locker, J Stevenson), G Der, JK Wing, and others*

General development of SCAN system and relevant training, coordination, data management and analysis: MRC Social Psychiatry Unit, Institute of Psychiatry, London

Chief Editor: *JK Wing*

SCAN Version 2 (1993 - 1994)

An Editorial Committee supervised changes necessary to incorporate the final revision of ICD-10 DCR and DSM-IV: *JK Wing (Chair), T Babor, PE Bebbington, A Bertelsen, TS Brugha, S Chatterji, W Compton III, G Harrison, V Mavreas, AJ Romanoski, N Sartorius, AY Tien and at WHO: TB Üstün, A Janca*

Textual changes were edited by T Brugha, S Chatterji, RY Mehta, AJ Romanoski, AY Tien, D Tien and M Tseng.

S Chatterji, G Der, AY Tien, and *TB Üstün* designed and produced new computer programs for data entry, CAPSE, and ICD-10 and DSM diagnostic algorithms. Inventa, Inc. of Bangalore, India (Principal Programmers: *R Ashok, CP Hari, and others*) undertook the programming. The programs have been tested at numerous SCAN sites. *SM Channabasavanna, RS Murthy*, and *MK Isaac* at the National Institute of Mental Health and Neurosciences in Bangalore, India, provided generous support for editorial and programming work.

Administrative and secretarial assistance: *R. Barrelet, J Head* (London), *G. Covino, J Wilson* (Geneva)

There is a list of SCAN Training Centers at the end of the Glossary in the Appendix.

INTRODUCTION TO SCAN

The Joint Project on Diagnosis (see acknowledgements) set up a Task Force on Diagnostic Instruments in 1980. This group decided to develop further two of the then most widely used instruments in epidemiological research. The NIMH-DIS, version III (with some input from PSE9, see below) formed the basis for the development and field trials of a new instrument, the Composite International Diagnostic Interview (CIDI), which has now been published by WHO. SCAN (see below) was the second instrument to be commissioned. The Task Force also commissioned a new instrument, the International Personality Disorder Examination (IPDE), which is nearing the end of its field trials.

SCAN (Schedules for Clinical Assessment in Neuropsychiatry) represents the latest stage in a 30-year line of development that began in the late 1950s. PSE9 was the first of the series to be published, following 15 years of work on earlier versions, including two large multicenter international projects – the US-UK Diagnostic Project and the International Pilot Study of Schizophrenia. PSE9 consists of only 140 items, compared to the 500–600 of PSE7 and PSE8. It has been widely used, as evidenced by its translation into 35 or more languages, but many users have regretted that the longer preceding editions were withdrawn; they would have preferred a choice, which is now provided by PSE10.

SCAN is a set of instruments aimed at assessing, measuring and classifying the psychopathology and behavior associated with the major psychiatric syndromes of adult life. It has 4 components: the tenth edition of the Present State Examination (PSE10), the Glossary of Differential Definitions, the Item Group Checklist (IGC) and the Clinical History Schedule (CHS). PSE10 itself has two parts. Part One covers somatoform, dissociative, anxiety, depressive and bipolar disorders, and problems associated with basic bodily functions and use of alcohol and other substance use. There is also a screen for Part Two conditions. Part Two covers psychotic and cognitive disorders and observed abnormalities of speech, affect and behavior.

Data from the schedules can be recorded in a variety of ways: on the SCAN Schedules themselves, on Coding Booklets, on the free-entry Record Booklet, into the computer data entry program, or by employing of the CASE computer program to administer the SCAN.

A set of computer algorithms (CATEGO-5) is used to process data entered from SCAN schedules. The output is presented as a series of options, including a range of profiles of symptom and IGC scores, an Index of Definition, ICD-10 and DSM-III-R categories, and a prediagnostic profile of categories. SCAN Version 1 items can also be derived from Version 2 of SCAN.

Further details concerning the development of SCAN, instructions as to its use, and the results of international trials of feasibility and reliability, will be found in the

Glossary, Training Pack and Instruction Manual. In its complete form, SCAN is intended for use only by clinicians with an adequate knowledge of psychopathology who have taken a course at a WHO-designated training centre. A shortened version of Part One of PSE10 can be used (e.g. in two-stage population surveys, as with the equivalent version of PSE9) by lay interviewers trained in these centers. All interviewers using SCAN should be fully familiar with the Glossary.

0 SCAN face sheet

0.001 ***Version of SCAN.***

0.002 ***Project/Centre number.***

0.003 ***Respondent number.***

0.004 ***Rater number.***

0.005 ***Key date of first PSE examination*** [Day Month Year]

D M Y

[If no examination, date of completion of first schedule]

SKIP => to 0.016 if non-routine option is chosen.

Only two periods may be rated using the routine option using the paper version of SCAN, e.g. a PSE and/or IGC for each period. If the computer version (CASE) is used, the onset and offset of individual SCAN items may be dated individually, and as many periods as needed can be rated.

Period(s) rated using PSE(s) - any source of information.

0 PSE not rated.
1 PSE rated for the period stated.

0.006 ***Present state (PS).***

0.007 ***Representative previous episode (RE).***

0.008 ***Lifetime before PS (LB).***

0.009 ***This PSE completed by***

0 PSE not used (IGC used instead).
1 Interviewer rating own interview.
2 Observer rating interview.
3 Rating from videotape.
4 Rating from audiotape.
5 Other, Specify ....

0.010 ***Source of information used to rate PSE.***

[Use left box for PS, right box for RE or LB.]

0 PSE not used [Only IGCs].
1 Respondent only.
2 Respondent and other sources.
3 Case records only.
4 Informant/s only.
5 Case records and informant/s. Specify relative, professional, etc...
6 Prepared abstract...

Period rated using IGC(s)

0 IGC not used.
1 IGC used for the period stated.

0.011 ***Present state (PS).***

0.012 ***Representative previous episode (RE).***

0.013 ***Lifetime before (LB).***

0.014 ***Source of information used to rate IGC.***

[Use left box for PS, right box for RE or LB.]

0 IGC not used.
1 Respondent only [e.g. observed behaviour ratings only].
2 Respondent and other sources.
3 Case records only.
4 Informant/s only.
5 Case records and informant/s. Specify relative, professional, etc................
6 Prepared abstract.......................

Clinical History Schedule (CHS)

[Use of CHS is recommended if a narrative output is required.]

0.015 ***Use of Clinical History Schedule.***

0 Not used.
1 CHS completed.

Use of the non-routine option (See also Special Episode list at items 1.019 - 1.024).

See Glossary. Employment of the non-routine option restricts or precludes the use of currently available SCAN computer algorithms and may hinder comparisons with other SCAN data sets.

0.016 ***Identifying number of period rated.***

0 Not used.
1-6 Number of episode rated in this PSE schedule.

Indicate the format for recording SCAN data (rate one option only).

0 Not used.
1 Used.

0.017 ***Rating boxes in SCAN schedules [PSE, IGC, CHS].***

0.018 ***Coding Booklet (fixed format).***

0.019 ***Record Booklet (free entry).***

0.020 ***Computer.***

SOCIODEMOGRAPHIC ITEMS

If there is any possibility of cognitive impairment, the following items may be completed as a test, by asking Respondent (R) for the information. Skip to Section 21 if necessary.

[CHS has a more comprehensive social history, which can be completed with R.]

0.021 ***Age*** [If not known, leave blank]

0.022 ***Date of birth*** [Day Month Year]

D M Y

0.023 ***Sex:***

1 Male.
2 Female.

0.024 ***Marital status:***

1 Married.
2 Cohabiting.
3 Separated.
4 Divorced.
5 Widowed.
6 Never married and not cohabiting.

0.025 ***Number of children born alive.***

0.026 ***Years of education.***

Number of years of formal education completed, not including vocational training, apprenticeships, etc. 98 = NK, 99 = NA.

0.027 ***Occupational status:***

1 Full time paid.
2 Part time paid.
3 Household duties.
4 Unemployed.
5 Retired.
6 Sheltered work.
7 Primarily a student.
8 NK.
9 NA (e.g. maternity leave).

0.028 ***Nature of paid employment (specify using local codes):................***

0.029 ***Number of months of paid employment worked past year.***

98 = NK, 99 = NA.

1 Beginning the interview

The purpose of Section 1 is to obtain a preliminary overview of the possible problems that the respondent (R) may have. Almost every item recorded in Section 1 can be rated with detail and precision in later Sections. The interviewer should be well informed before starting the interview and be prepared for the following possibilities.

In general

Begin with Section 1 and then proceed to those sections containing items that R most wants to discuss or that are clearly predominant. This will provide more complete coverage on the important topics. If the interview is likely to be incomplete, try to complete sections with highly relevant items first.

Consider the kind of interview likely to be needed.

Prior to beginning, the interviewer should be as well informed as possible. Usually there will be information will be available as to whether R is likely to be able to answer questions and give a fair account of any problems. Several contingencies are outlined below.

If none of the following contingencies apply, begin the interview at next page.

CONTINGENCIES

1 Severe language disorders/or cognitive impairment

- Rate Section 15 for language disorders.
- Rate Items **21.089 - 21.102** (behavioral and history items).
- Rate Sections 22 - 25.
- Complete Clinical History Schedule.

2 Severe behavior disturbance, uncooperativeness or likelihood of premature termination

Begin with those Sections that are most relevant to R. Keep conversation going while observing speech, affect and behavior (Sections 22 - 25). If necessary complete the examination in stages. All stages can be rated on one schedule if the full SCAN can be is completed within a few days. Be sure to rate adequacy items (**13.129** and **20.092**).

- Re-interview with PSE10 as opportunity affords.
- If interview impossible, rate PS on Item Group Checklist.
- Use all information available to rate Item Group Checklist for previous episode if necessary.
- Complete Clinical History Schedule.

3 Recent catastrophic trauma or psychosocial stressor

- Begin with Section 13 and complete required checklist items if needed.
- Use informant/records/standardized inventories to check details of event.
- Complete the full SCAN as soon as possible, and make attributions of effects of stressors in Section 13, if possible, delaying until precise "causal" influence of any trauma/stressor is clear.

4 Dissociative symptoms

- Complete items **2.053 - 2.074**.
- Obtain information from informants/records.
- Complete the full PSE10.

5 Drug or alcohol use

If the main problem, it is advisable to consider Sections 11/12 first, but the whole of PSE10 must be completed. Attributions of cause and effect may be rated in Section 13, preferably after their influence has been clearly established.

6 Eating disorders

- If the main problem, begin with Sections 8 - 9.
- Return to Section 1 and complete PSE.

INTRODUCTIONS AND OVERVIEW OF NEUROTIC, PSYCHOTIC AND SUBSTANCE USE PROBLEMS

The Clinical History Schedule contains a more comprehensive clinical and social history, which may be completed with the additional aid of case-records and informants. There is a reminder in CHS to check all PSE history items in the light of information from case records and informants.

The Glossary notes on rating episodes should be used when completing this section.

Always be as fully informed as possible, e.g. from case-records and/or other informants before starting.

My name is [I am] ... What is your name?
[Offer to shake hands]

Obtain permission to proceed and to use any recording equipment. Explain the purpose of the interview, for example:

We are making a survey in this neighborhood in order to:

Or:

This is part of the routine medical examination of people who come to the hospital, except that we should like to take time to discuss problems in more detail than usual because ...

Or:

We are making a study of ... and would like to ask you about any health problems you have been having

Or:

[Other introduction and explanation]

Is that all right with you?

To begin with, I should like to get a general view of the kind of health problems, physical and mental, that might have troubled you recently. Would you say that you had recently enjoyed good mental and physical health or have you had any problems?

Allow or prompt R to give a narrative account. Ask clarifying questions only, as in the following prompts:

- *Could you describe what ... was like?*
- *Could you give me an example of ...?*
- *Do you still ...?*
- *When did ... start?*
- *Are you taking any medication? (What kind?)*
- *Are you receiving any other help or therapy?*

If R describes no relevant symptoms, ask:

Would you say you had enjoyed excellent physical and mental health recently, with no real problems?

1.001 ***R's unprompted response to initial questions*** ☐

0 R does not describe having had psychotic, affective, or neurotic symptoms recently.
1 R spontaneously describes such symptoms.
2 R reluctant to describe symptoms but does so after probing.
8 R's reply is uninformative.

Use all information from other sources to rate whether the following symptom types have ever been present:

[Use colloquial terms. Do not ask questions that R has already answered, but confirm them, e.g. 'So you have had ...']

Were specified symptom groups ever present?

May I mention some problems that many people experience at some time in their lives, and ask whether you have ***ever*** *been troubled by any of them, and if so how distressing or disabling they were?*

For each type of symptom ever present, ask:

- *Could you describe what it was like?*
- *When did it first start?*
- *How old were you then?*
- *Has it been continuous or intermittent?*
- *And how long has this recent period lasted?*

0 No evidence that symptom category was ever present.
1 Present, but only mildly distressing or disabling.
2 Present and moderately distressing or disabling.
3 Present and severely distressing or disabling.

1.002 ***Anxious or panicky feelings, often in particular situations?*** ☐

1.003 ***Feeling very low in spirits?*** ☐

1.004 ***Feeling the opposite - feeling much too 'high' or elated without reason?*** ☐

1.005 ***Having experiences that are difficult to explain or understand?*** ☐

1.006 ***Having problems due to alcohol or other substances?*** ☐

CUT OFF => to 2.001 and begin PS-Check if 1.002 - 1.006 are all rated 0.

SELECTION OF EPISODES

Decide provisionally, on the basis of this information, which periods represent the most clinically significant symptoms (of neurotic, affective or psychotic disorders) during the clinical course. See Glossary notes on choice of episodes. Time periods of symptoms that do not fit the pattern established here in Section 1 can be recorded Separately in each Section.

1.007 ***Onset of Present Episode, PE*** [month, year] ☐☐☐☐

Onset date must follow 3 months or more without significant clinical symptoms. R must have had clinically significant symptoms (any severity level) within approximately the month before the interview. See notes for full definition. If not 'in episode' at interview, leave blank, but complete a PS-Check for 4 weeks before examination. September, 1994 is entered 0994.

1.008 *Number of days in PS* ☐☐

This period should usually be 28 days and end at the date of examination. See Glossary notes for permissible extensions, e.g. 'notional month'.

1.009 *Dates of Representative Episode, RE* ☐☐☐☐ to ☐☐☐☐

Use RE if a single discrete episode, together with PS, adequately represents the clinically significant symptoms manifested during the course. See Glossary for full definition. March 1971 to May 1973 is entered as 0371 to 0573. If RE is contingencies with the beginning of PS, PS+RE=PE. Variations in the timing of RE between different syndromes, if necessary can be recorded separately within each Section.

SKIP => to 1.014 **if RE used. If RE not used, leave RE dates blank, and consider 1.010 - 1.013. Record the option chosen for PERIOD (PS, RE, LB) at 0.006 - 0.008.**

Dates of Lifetime Before, LB

LB is dated from the onset of the first episode to the start of PS. LB can represent one continuous episode or contain several discrete episodes. Use LB if PS and RE together do not adequately represent the clinically significant symptoms manifested during the course. It would not usually be sensible to rate more than 2 episodes in LB if R is also in episode at PS. See Glossary for full definition and rating instructions. Leave blank if never present. Sections not referred to here are rated for specific periods referred to above(contingencies), i.e. Sections **2, 9, 11, 12** and **21**.

1.010 *Episode of Neurotic symptoms* ☐☐☐☐ to ☐☐☐☐

1.011 *Episode of Depressive symptoms* ☐☐☐☐ to ☐☐☐☐

1.012 *Episode of Manic symptoms* ☐☐☐☐ to ☐☐☐☐

1.013 *Episode of Psychotic symptoms* ☐☐☐☐ to ☐☐☐☐

Date approximate age of first onset of symptom types in LB

This estimate should represent the first onset of a clinically significant group of symptoms of the type listed, causing at least moderate distress or disability. 98 = NK. 99 = NA.

1.014 ***Age at first onset of Neurotic symptoms***

1.015 ***Age at first onset of Depressive symptoms***

1.016 ***Age at first onset of Manic symptoms***

1.017 ***Age at first onset of Psychotic symptoms***

1.018 ***Quality of remissions between episodes***

If there have been 2 or more episodes, whether similar or dissimilar, consider the persistence of significant clinical symptoms between them.

0 No episode symptoms.
1 No or few significant symptoms between episodes.
2 Mixed, but intervals more often like (1).
3 Mixed, but intervals more often like (4).
4 Significant symptoms usually present in intervals.
5 Continues symptomatology

SKIP => to 1.025 if episode(s) have been rated at 1.010 - 1.018.

Special episode list

Use of this episode list entails rating several episodes separately, each with its own face sheet and recording book. Item **0.016** of the face sheet must contain the identifying number of the episode rated. Any other information needed locally should be entered in the 'spare' spaces provided in the data entry program. A separate CATEGO outputs will be provided for each individual episode. Users will make their own analyses of the results. The dates of recurrent episodes of affective disorder may also be recorded in a similar way in Sections 6 and 10.

1.019 ***Episode 1*** to

1.020 ***Episode 2*** to

1.021 ***Episode 3*** to

1.022 ***Episode 4*** to

1.023 ***Episode 5*** to

1.024 ***Episode 6*** ☐☐☐☐ to ☐☐☐☐

MEDICATION AT THE TIME OF INTERVIEW

You mentioned that you were taking medication on prescription.

- *Do you know what kind it is?*
- *Who prescribed it? What effect does it have?*
- *Has it helped the ... symptoms?*
- *Are there any unpleasant effects? What are they like?*
- *What are they like?*

Rate each type of medication.

0 Not used at the time of interview.
1 Used, but unlikely to have altered symptoms.
2 Used and may alter presence/severity of symptoms (consider individual items etiology option for individual items or optional rating of attribution at Section 13).
8 NK.

1.025 *Sleeping pills* ☐

1.026 *Anti-anxiety* ☐

1.027 *Anti-depressive* ☐

1.028 *Anti-manic* ☐

1.029 *Anti-psychotic* ☐

1.030 *Anti-Parkinson* ☐

1.031 *Anti-alcohol* ☐

1.032 *Analgesics* ☐

1.033 *Other medication - Specify*............................... ☐

1.034 *Unidentified medication* ☐

1.035 ***Global Improvement from medication by time of examination*** ☐

0 No apparent effect.
1 Moderate improvement in symptoms.
2 Substantial improvement.

1.036 ***Effects of medication on interview***

0 None seen.
1 Some problem from side effects but not severe.
2 Severe side effect.
3 Marked interference with interview.

It may be necessary to make further detailed attributions of effects of medication after completing SCAN (i.e. at Sections 13 and 20).

PSYCHOSOCIAL INTERVENTIONS

1.037 ***Psychosocial intervention at the time of interview***

You mentioned that you were receiving help/support/counseling.

- *Do you know what kind it is?*
- *Who provided it? What effect does it have?*
- *Has it helped the ... symptoms?*
- *Are there any upsetting or unpleasant effects? What are they like?*

Rate psychosocial interventions at time of interview.

0 Not used at the time of interview.
1 Used, but unlikely to have altered symptoms
2 Used and may alter presence/severity of symptoms (consider optional rating of attribution at Section 13).
8 NK.

1.038 ***Formal psychotherapy***

1.039 ***Crisis counselling***

1.040 ***Other formal intervention***

1.041 ***Informal/family or friendship support***

1.042 ***Other form of support/assistance:***

Specify ....

1.043 ***Global Improvement from psychosocial intervention by time of examination***

0 No apparent effect.
1 Moderate improvement in symptoms.
2 Substantial improvement.

It may be necessary to make further detailed attributions of effects of psychosocial intervention/support after completing SCAN Part One (Section 13).

OTHER ASPECTS OF CLINICAL HISTORY

A more complete record should be made in CHS.

1.044 ***Family history in parents, sibs, children***

0 No.
1 Yes, symptoms similar in type to Respondent's.
2 Yes, but dissimilar symptoms.
8 NK.

Entries for items **1.045 - 1.047** should be checked when completing the CHS.

1.045 ***Age in years at first prodromal symptoms***

Irrespective of type of symptoms (see Glossary).

98 NK.
99 NA.

1.046 ***Age in years at first clear recognition of disorder [1.045] by R or others***

1.047 ***Age in years at first contact for formal advice***

1.048 ***Speed of onset of first episode of any symptoms***

1 Within a month.
2 Within 1-3 months.
3 Within 3-12 months.
4 More than a year.
8 NK.

1.049 ***Social competence before recognized onset***

0 Normal within culture.
1 Moderate degree of social disablement.
2 Marked and noticeable social disablement.

1.050 ***Total number of admissions to hospital for psychiatric disorder***

98 NK.
99 NA

1.051 ***Length (months) of longest period in hospital***

97 97 or more months
98 NK
99 NA

1.052 ***Number of previous episodes of clinically significant symptoms***

Rate as at **1.049**; dates of previous episodes of affective disorder can be recorded in Sections 6 and 10; certain aspects of previous episodes of psychotic disorders are recorded in Section 20.

Rating scales in Part One

GENERAL POINTS

The main rating scale in Part One is Scale I. The rating points are specified in the SCAN text, at the beginning of Section 3, and elaborated on the next page of the Glossary.

However, many items (particularly in other sections of Part One) have their own individual rating scales, which are specified within their item blocks. It should be noted that all symptom items can be rated at points 0, and 5-9, using the definitions given in Scale I, whether or not these points are specified in the text.

For making etiological attributions, many items have boxes with dashed line below the episode rating boxes. These are optional ratings that are not required, but which may be useful for research purposes, and in certain clinical situations. The purpose is to separate the ratings of phenomenology from any ratable etiological factors; for example Parkinson's disease, or the effects of alcohol or other drugs. The etiology scale should be used to code any attribution of etiology.

RATING SEVERITY

The severity of a symptom can be assessed in terms of duration, persistence, degree of interference with other mental functions, distress, impairment of everyday activities, effect on other people, and contact with services of various kinds. In SCAN, the approach is to measure clinical severity by the duration and frequency of the symptom and the degree of interference with mental functions (intensity). Social and occupational performance, other people's reactions, and help-seeking behavior (all of which can be influenced by many other factors), are assessed separately.

Somewhere between the two lies the respondent's own reaction, but this may be stoical or distressed according to temperament and circumstances, and is therefore also regarded as adding a degree of ambiguity to the rating. Distress is only mentioned in items which use criteria from a rule-based system that requires it to be present.

These points hold good for ratings throughout SCAN and, in particular, for the four main Rating Scales.

RATING SCALE I

Many items in Sections 3-6 are rated on this standard 4-point scale (0-3). The factors chiefly involved are intensity (intrusiveness and extent of interference with mental functions) and frequency of the symptoms. The standard definitions are suitable for a period of about 4 to 6 weeks. This is the period of the 'Present State', from which the PSE originally derived its name.

In longer episodes, it is often possible to select (with R) a period of equivalent length during which most of the symptoms characteristic of the episode were present (a 'representative month').

When rating a period longer than a month it may be difficult to specify the proportion of time during which a symptom has been present.

In such cases, rate mainly on clinical intensity rather than duration. Frequency can still, however, be used as appropriate.

Do not be too pedantic when asking about each symptom. It is usually sufficient to establish an overall frequency and intensity for a group of symptoms and then to establish any variation in particular items.

0. This is a positive rating of absence. It does not mean 'not known' or 'uncertain whether present or not'. It can only be used if sufficient information is available to establish its accuracy.

1. This is a positive rating of presence, but presence of such a minor degree that it is not appropriate for use in classification. Like (0), it does not mean 'not known' or 'uncertain'. Ratings of (1) count in scores, which in turn influence the level allocated on the Index of Definition.

2. This rating means that the item is present at a level sufficient to use in classification. For this purpose it is equivalent to 3, but it contributes less to scores. In general, it is used when symptoms are of moderate severity during most of the period being assessed.

3. A rating of (3) is similar to (2) except that the symptom is present in severe form for most of the period under review.

5. The presence of psychotic symptoms can make the rating of Part One items very difficult, because of problems in interpreting the meaning of what R says, or because the symptoms (for example, anxiety or a phobia about leaving one's house) may themselves be based in psychotic experiences. The rating should only be made when there is genuine doubt about the nature of the symptom or the balance is in favor of the symptom being psychotic.

8. If, after an adequate examination, the interviewer is still not sure whether a symptom is present (rated 1-3) or absent (rated 0), the rating is (8). This is the only circumstance in which (8) is used. It should not be used to indicate a mild form of the symptom.

9. This rating is only used if the information needed to rate an item is incomplete in some respect, for example because of language or cognitive disorder, or lack of cooperation, or because the interviewer forgot to probe sufficiently deeply. It

6 Physical problems impair rating
7 Trait

is distinguished from (8) because the examination was not, for whatever reason, carried out adequately.

In the SCAN text, an instruction to 'use Scale I' simply means that it is not necessary to point out any individual rating characteristics for that item. Any point on Scale I can be selected, according to clinical judgement.

OPTIONAL RATINGS OF ITEM-SPECIFIC ATTRIBUTIONS OF CAUSES, AND TRAIT RATINGS

SCAN users and raters with qualification and practice experience in medicine and clinical pharmacology, studying patients following appropriate, extensive, physical and laboratory investigation may wish to rate judgements of definite attributions of cause at an item level. For each period first rate the item using the appropriate SCAN rating scale (I-IV), in the standard scoring boxes, and then the nature of the attribution in the dashed boxes below the standard boxes, using the scale provided. Up to two periods can be rated. Note that the conventional use of 0,8, and 9 does not apply and these scores have specific meaning. The items themselves already should have been rated purely on the basis of certainty, presence, and severity.

The following optional ratings should only be made by investigators planning to study their validity. The reliability of making these ratings should be established. They represent a more detailed way of recording ratings of attribution of organic cause throughout the SCAN. There is also an opportunity to make a trait rating of individual items (i.e. a rating of "7" on rating SCALE I of SCAN version 1). Users should contact their training center for information of employment in SCAN algorithms.

ATTRIBUTIONAL RATING SCALE (Optional)

Use the attributional rating scale to rate nature of influence on symptom presence or severity at the item level. The rating of 0 is the only 'negative' rating, and can be used if there is clear evidence that treatment has resulted in a reduced rating. Ratings 1 - 8 represent attributions of etiologic effect, increasing the likelihood of an item being rated or its severity being increased. These attributions must only be made for symptoms/items that have positively rated, and should be entered into the dashed boxes below the standard episode rating boxes. The rating of 9 is reserved for indicating a trait characteristic. If no effect or trait is judged to be present leave the dashed box blank.

0 Item rating reduced by effects of treatment
1 Alcohol
2 Other psychoactive substance
3 Effects of somatic psychiatric treatments (electroconvulsive therapy, antidepressant medication, neuroleptics, etc)
4 Known primary intracranial process (Alzheimer's disease, Huntington's disease, Parkinson's disease, tumor, stroke, etc.)
5 Non-psychiatric medication, toxins
6 Other medication condition 1 (specify at Section 13 and/or Section 20)
7 Other medication condition 2 (specify at Section 13 and/or Section 20)
8 Other medication condition 3 (specify at Section 13 and/or Section 20)
9 Trait. Essentially lifelong characteristic of R.

2 Physical health, somatoform, and dissociative symptoms

Adapt questions about disorders and treatments to local usage. Use information from informants and case records.

2.001 ***Physical fitness, past month***

Now I should like to ask you some questions about your physical health. ***During the past month or so*** *would you say your physical health has been excellent, good, fair or poor.*

1 Excellent.
2 Good.
3 Fair.
4 Poor.

If fair: *What makes you say 'fair' rather than 'good'?*

Continue with **2.002** if necessary.

If poor/fair:

2.002 ***Length of unfitness***

For how long has your physical health been poor?

Specify in years and months. One year and 6 months is entered '0106'.
Leave blank if physical health has been at least fair.

2.003 ***Change in weight, past year***

Have you lost or gained any weight during the past year? [1 kg = 2.2 lb].

- *Over what period of time?*
- *What is the reason for the change?*

0 Virtually no change.
1 Less than 2 kg (<5 lb) change.
2 Lost 2+kg (5+lb) within 6 week period.
3 Gained 2+kg (5+lb) within 6 week period.
4 Weight has fluctuated, outside 2+kg (5+lb) loss or gain.

Items **8.006** and **8.007** deal with weight change in more detail.

2.004 ***Physical illnesses or disabilities, past year*** ☐

In the past year, have you had any bodily aches or pains, or weakness, or physical illnesses or injuries, or disabilities that limited your activities in any way?

- *What sort of problems have you had?*
- *Did the doctor/specialist tell you what was the matter?*

0 None.
1 Describes somatic symptoms or disorder without clear diagnosis.

SKIP => to 2.007 or 2.008

2 Has clear physical diagnosis.

DIAGNOSABLE PHYSICAL ILLNESS OR DISABILITY

Conditions associated with mental retardation and/or autism, e.g. Down's Syndrome, fetal rubella, etc., should be checked in CHS. The first box should contain a letter indicating the ICD-10 chapter, followed by up to 3 digits. If none leave blank.

2.005 .... ☐☐☐☐

2.006 .... ☐☐☐☐

Women only: [Show a card if it would be helpful.]

2.007 ***Pre-menstrual symptoms in the past year*** ☐

Have you had any symptoms before menstruation which stop as soon as the period starts? Such as:

Irritability,
Depression,
Feeling of being bloated or gaining weight,
Tenderness or swelling of the breasts,
Muscular tension,
Aches and pains such as headaches, backache, etc.,
Poor concentration,
Cravings.

Enter number positive; if 7 or 8 rate 7, 8 = NK. Ideally this rating should be based on a concurrent diary using a systematic inventory covering at least two cycles.

2.008 ***Limitations on physical activities, past year***

Have any of these problems ***(2.001 - 2.007)*** *limited your well-being or activities in the past year?*

- *How severe has the limitation on your physical activities been?*

0 None.
1 Mild.
2 Moderate.
3 Severe to incapacitating.

2.009 ***Satisfaction with care, past year***

Have you had expert advice about ... (physical problems)?

- *How much contact have you had with doctors [healers, etc.]?*
- *What did the doctor say was wrong?*
- *Are you reasonably satisfied that the problem has been thoroughly investigated and you have received good advice?*
- *Have you any more appointments with a doctor?*

0 R is satisfied that the problem has been investigated and treated with reasonable care.
1 R says that doctors think nothing is wrong or do not take the problems seriously.
8 Unclear whether dissatisfaction is due to R's overconcern or to reasonable grievance.

CUT OFF => to 3.001 if there is no evidence from interview, records, or informants of somatoform or dissociative symptoms.

Always continue if symptoms have no convincing medical explanation.

SOMATOFORM SYMPTOMS

The PS rated in the rest of this section should include the past two years.

Rate any probable contributions to somatoform or dissociative symptoms from subclinical infections or nutritional problems, etc., at **2.079**.

Use medical records, if available, to rate the following items.

Review R's description of symptoms and ask for further details.

I should like to be sure which physical symptoms you have experienced. You mentioned ... What was that like?

- *Did you see a doctor about this?*
- *Did you have special tests or X-rays?*
- *Did you have any treatment? What kind?*
- *What did the doctor tell you was the matter?*
- *What did you think about it? Were you satisfied?*

Complete 2.010 - 2.035 with further probes as necessary. Use a prompt card if it would be helpful.

Consider local expression of these symptoms, e.g. loss of semen would be rated at **2.023**. If a more detailed list is required, allowing each symptom to be rated separately and containing additional items, complete checklist at the end of Section 2, items **2.080 - 2.135**, and then continue at **2.037**. Ask for details if descriptions are unclear. See Glossary for definitions.

The checklist includes certain autonomic symptoms, also found in anxiety disorders. If such symptoms occur in the context of panic attacks, rate relation to panic attacks at **2.136** and complete Section 4 fully. In general, physical disease should already have been excluded; rate 0 or 9 (not applicable) if any symptom has an adequate physical explanation. In general for a symptom to be rated as present it should NOT be medically explained and should be distressing to the R. Somatoform complaints usually begin before the age of 30, and later complaints often turn out to have a physical/medical explanation.

0 Symptom absent.
1 Symptom present. No full medical explanation.
8 Uncertain whether symptom is present.

2.010 ***Shortness of breath when not exercising.*** ☐☐

2.011 ***Chest or heart pain, discomfort, burning.*** ☐☐

2.012 ***Vomiting, regurgitation, aerophagy, hiccough.*** ☐☐

2.013 ***Difficulty with swallowing.*** ☐☐

2.014 ***Stomach or bowel pain, wind, indigestion, increased bowel movements.*** ☐☐

2.015 ***Problems with passing water, frequency.***

2.016 ***Pain in arms or limbs, joints, hands or feet.***

2.017 ***Any other aches or pains, headache.***

2.018 ***Loss of memory.***

2.019 ***Any problem with sight, hearing or speech.***

2.020 ***Convulsions, trances, faints, dizziness.***

2.021 ***Problems with movement, paralysis.***

2.022 ***Numbness, loss of sensation, tingling.***

2.023 ***Genital problems, pain, discomfort, sensations, menstrual problems, sexual indifference.***

2.024 ***Blotchy or discolored skin (not Rosacea).***

2.025 ***Palpitations, pounding heart.***

2.026 ***Hot or cold sweats.***

2.027 ***Dry mouth.***

2.028 ***Flushing or blushing.***

2.029 ***Churning or discomfort in stomach.***

2.030 ***Other pain.***

2.031 ***Nausea.***

2.032 ***Unusual or copious vaginal discharge.***

2.033 ***Bad taste in mouth, excessively coated tongue.***

2.034 ***Frequent, loose bowels or fluid discharge from anus.***

2.035 ***Other symptoms, specify:* ....**

2.036 ***Length of distress due to preoccupation*** (months)

To summarize then, you have symptoms of ...
Have you been worried that you have a serious illness(es)?

- *Have you been distressed by these worries?*
- *How long have you felt this way? How old when it first started?*

000 Absent or less than 1 month.

2.037 ***Variability of somatoform symptoms***

Have the symptoms changed much during that time or have they stayed more or less the same throughout?

0 Little change.
1 Some but not frequent changes in symptoms.
2 Frequently changing symptoms.

2.038 ***Hypochondriacal preoccupation***

How many doctors have you consulted in the past 2 years?

- *What investigations were made? With what results?*
- *Were the doctors/healers reassuring?*
- *Why do you think there is something physically wrong?*
- *Have you been told the complaint is nervous?*
- *Does that seem likely to you?*
- *Have you been dosing yourself with any medications?*

Consider the degree of pressure, preoccupation, repetition and distress expressed in the complaints, frequency of seeking a 'medical' explanation in spite of reassurance.

Rate the length of preoccupation at **2.036**.

0 Condition not present.
1 Present but mitigated for a time by reassurance (i.e. for a few weeks during or after investigations have been carried out).
2 Present and apparently unmodifiable.

2.039 ***Hypochondriacal conviction***

What do you yourself think is wrong?

- *Do you think you have a serious disease?*
- *Or more than one? (Are you upset about it?)*
- *Have you told the doctors about this? (What do they say?)*
- *Did you feel relieved when they said nothing was wrong?*
- *Have you still been convinced, in spite of the experts?*
- *What medications do you take? (On doctor's prescription?)*

R expresses delusional conviction in face of evidence to the contrary, that he or she has one or more serious diseases. Exclude fear of contracting a disease rated at **4.044**. Include Dysmorphophobia, and rate also at **16.011**. Include hypochondriacal delusions in context of depression, rated at **6.020 or 19.027**.

0 R does not attribute symptoms to one or two particular diseases.
1 R names only 1 or 2 specific diseases or deformity as causing the symptoms, and belief must continue for at least 6 months duration.

Rate items **2.040 - 2.049**:

0 Absent.
1 Present.

2.040 ***Excessive and distressing fatigue (physical exercise)***

Do you become exhausted and weak after taking even minor exercise?

- *Is it distressing? Can you describe what happens?*
- *Do you get muscular aches and pains afterwards?*
- *What about other unpleasant bodily feelings?*
- *Did it follow any particular illness?*

Rate irritability if present at **3.009.**

2.041 ***Excessive and distressing fatigue (mental exercise)***

Have you had that kind of fatigue after mental effort, for example, reading or other kind of mental activity?

- *Is a distressing effort to concentrate your attention on anything?*

2.042 ***Inability to recover normally from fatigue***

If you get these symptoms of mental and/or physical fatigue, how long does it last?

- *Can you recover by rest or relaxation?*

If 1 or more symptoms [items **2.010 - 2.035**] occur in any of the following systems:

Do you think there is something wrong with ...?

Rate preoccupation with symptoms within a bodily system or structure (more than one of the following may be rated if present).

2.043 ***Heart and cardiovascular***

Da Costa syndrome, cardiac neurosis, neurocirculatory asthenia.

2.044 ***Oesophagus and stomach***

Psychogenic aerophagy, hiccough, gastric neurosis, wind in stomach.

2.045 ***Lower G.I. tract***

Psychogenic irritable bowel syndrome, psychogenic diarrhoea, gas syndrome.

2.046 ***Respiratory system***

Hyperventilation, etc.

2.047 ***Urogenital system***

Psychogenic frequency, psychogenic dysuria, etc.

2.048 ***Other psychosomatic problem***

If pain is rated as present in items **2.010 - 2.035** rate **2.049** and **2.050**.

2.049 ***Pain syndrome***

Could you tell me more about the pain you mentioned earlier

- *How long does it last?*
- *What were the circumstances in which it began?*

E.g. traumatic events, disturbed relationships, insoluble problems, etc.

2.050 ***Pain syndrome duration***

Complaints of persistent, severe and distressing pain not explained or exacerbated by any physiological process or disorder.

Enter duration in months.

000 Absent.

2.051 ***Elaboration of physical symptoms***

Symptoms originally due to a physical disorder but exaggerated or prolonged by psychological factors - compensation, fear of death, hope for more successful treatment, etc. Check at **26.058**.

0 Absent.
1 Present.

2.052 ***Factitious disorder***

Intentional production or feigning of physical or psychological symptoms or self-infliction of wounds. External motivation for elaborating symptoms as in **2.051**. Include Munchausen's syndrome: check at **26.057**. If there is such motivation, i.e. malingering, code at Z 76.5 in ICD-10.

0 Not present.
1 Physical but not psychological symptoms present.
2 Psychological but not physical symptoms present.
3 Both psychological and physical symptoms present.

DISSOCIATIVE SYMPTOMS

The subsection on somatoform disorders (items **2.010 - 2.035**) contains most of the items needed to identify dissociative symptoms as well. Many of these items may require further assessment as follows.

If item **2.018** identifies memory problems, ask the following as appropriate:

2.053 ***Dissociative amnesia***

You mentioned that you had memory problems. Could you tell me more about them, please? Are they present all the time or do they come and go? (In what way?)

0 No memory problems, or only ordinary forgetfulness.
1 Partial and fluctuating amnesia.

2.054 ***Amnesia centered round recent stress***

When did the memory problems start?

- *Were you under any particular stress then?*
- *What were your personal relationships like at that time?*

E.g. traumatic events, disturbed relationships, insoluble problems, etc.

0 No memory problems.
1 Partial and fluctuating amnesia centered on stress.

2.055 ***Fugue associated with stress***

Have you ever found yourself a long way from your usual range of travel without being able to remember how you got there?

- *Were you under stress then?*

Apparently purposeful travel beyond usual range, not just 'wandering', with amnesia for the period. See **21.030.**

0 No evidence of dissociative fugue.
1 Evidence of dissociative fugue with amnesia associated with stress.

If memory problems are present, use Section 21, cognitive impairment.

If items **2.010 - 2.035** identify problems of loss of consciousness, ask the following questions as appropriate:

2.056 ***Dissociative stupor***

You mentioned that you had lost consciousness or experienced trances or fits, recently. Could you tell me more about that, please?

- *Did you seem to others to be completely motionless and unresponsive for long periods?*

If informant available: diminution or absence of voluntary movement, speech, responsiveness to light or noise; breathing and muscle tone normal; may be some co-ordinate eye movements. See also **2.057** and **2.070**.

0 No evidence of dissociative stupor.
1 Evidence of dissociative stupor.

Rate items **2.057 - 2.069:**

0 Absent.
1 Present.

2.057 ***Trance experience***

Have you experienced being in a trance?

- *Did you lose your sense of personal identity?*
- *Can you describe the experience?*
- *Did you want it to happen?*
- *Was it troublesome for you?*

If trance experience initially welcomed:

- *Did the trance continue without your wishing it?*
- *Did it start at a religious or social occasion?*

Rate trance only, whether or not there is a possession experience (see **2.058**): narrowing of awareness or selective focusing; limited movement; posture and speech restricted to small repertoire. The experience is unwanted, troublesome, and not part of, unless an unwanted prolongation of, a religious or culturally accepted ceremony or situation. See also **2.070**.

2.058 ***Possession experience***

Have you had the experience of being taken over by some other power?

- *By what? A spirit, deity, person?*
- *Did you lose your sense of personal identity?*
- *Can you describe the experience?*
- *Did you want it to happen?*
- *Was it troublesome for you?*

If possession experience initially welcomed:

- *Did it continue without your wishing it?*
- *Did it start off at a religious or social occasion?*
- *Have you had that possession experience without being in or going into a trance?*

Rate possession experience only, whether or not there is a trance. The experience is unwanted, troublesome, and not part of, unless an unwanted prolongation of, a religious or culturally accepted ceremony or situation.

2.059 ***Possession experience combined with trance***

Definitions as in **2.057** and **2.058**. One or other experience, or both, must be unwanted and troublesome. See also **2.064**.

2.060 ***Dissociative convulsions***

Have you had any faints or fits or convulsions recently?

- *What were they like?*
- *Did you lose consciousness altogether?*
- *Did you hurt yourself at all by falling or biting your tongue.*
- *Did you pass water?*

Sudden, spasmodic movements, not accompanied by loss of consciousness or clouding or harm to self, such as tongue biting, bruising or incontinence. No evidence of organic disorder.

2.061 ***Association of 2.056 - 2.060 with stress***

*When you experienced these problems [**2.056 - 2.060**] were you under any particular stress?*

- *What were your personal relationships like at the time?*

E.g. traumatic events, disturbed relationships, insoluble problems, etc.

SKIP => to 2.064 if somatoform list (items 2.010 - 2.035) identifies no problems of sensory loss.

2.062 ***Dissociative sensory loss or anaesthesia***

Have you had any loss of vision, hearing, vibration, smell or touch? Can you feel heat and cold normally?

The way the sensory deficits are manifested indicates a lack of medical knowledge, e.g. a glove or stocking distribution. Beware sophistication! See also **2.064** and **19.024** (culture specific disorders).

2.063 ***Association of 2.062 with stress***

When you experienced this problem were you under any particular stress?

- *What were your personal relationships like at the time?*

E.g. traumatic events, disturbed relationships, insoluble problems, etc.

If item **2.021** identifies problems of voluntary control of movement, ask **2.064** and **2.065**.

2.064 ***Dissociative disorder of voluntary movement***

Have you been unable to move an arm or a leg, or some part of a limb? Have you found your movements uncoordinated or that you could not stand unaided?

- *Or found that you could not speak?*
- *Can you tell me about that, please?*

2.065 ***Association of 2.064 with stress***

When you experienced this problem were you under any particular stress?

- *What were your personal relationships like at the time?*

E.g. traumatic events, disturbed relationships, insoluble problems, etc.

Multiple personality

Rate items **2.066 - 2.069:**

0 Absent.
1 Present.

2.066 ***Two or more discrete personalities***

2.067 ***Each personality is complete***

2.068 ***Each manifested for discrete periods during which one of them takes control of the persons behavior***

2.069 ***Extensive forgetfulness with inability to recall important personal information***

Rate dissociative hallucinations at **17.020.**

2.070 ***Other dissociative states***

Residual category.

All occur in circumstances suggestive of a psychogenic aetiology and there are usually other dissociative symptoms.

1 Transient amnesic or trance states of adolescence.
2 Ganser syndrome ("Past the point" answers, approximate answers suggesting that R is aware of the correct answer). [Check at item **23.010**].
3 Poorly differentiated multiple personality.
4 Depersonalization derealization disorder. [Check at item **3.012**].
5 Culture specific dissociation disorder, e.g. Latah, Piblokto, Ufufuyana, Saga (refer to Annex 2, ICD-10 DCR and item **19.024**). Specify:
6 Other, specify: ....

2.071 ***Age at onset of present somatoform disorder.***

2.072 ***Age at first onset of somatoform disorder.***

2.073 ***Age at onset of present dissociative disorder.***

2.074 ***Age at first onset of dissociative disorder.***

2.075 ***Dates of PERIOD of Section 2 symptoms*** to

It is necessary to record only if dates of symptoms in this Section are different from PERIOD/S rated in Section 1 (items **1.007, 1.008, 1.009**).

If PERIOD used is RE or LB, record start and finish dates of PERIOD of symptoms in this Section. If PS (and dates of PE) used record only start date of PE:

Onset of PS/PE, RE, LB (Left rating boxes)

Offset of RE or LB (Right rating boxes)

2.076 ***Interference with activities due to Section 2 symptoms***

You have mentioned [summaries symptoms] *during PERIOD. Overall, how much interference has there been with your everyday activities because of these problems?*

Rate interference due to symptoms in Section **2.**

0 No Section 2 symptoms present in significant degree.
1 Symptom/s present but little interference.
2 Moderate or intermittent interference.
3 Severe to incapacitating interference.

Although opportunity to rate attribution of physical (including alcohol, other drugs, etc.) cause is present using the optional attributional scale and the dashed boxes, items **2.077** and **2.088** allows the user to re-rate organic cause at the section/syndrome level.

2.077 ***Organic cause of any dissociative symptoms***

0 Absent.
1 Probable organic cause; not fully confirmed.
2 Definite; confirmed by expert investigation.
8 Uncertain whether organic or not.

2.078 ***Identify organic cause of dissociative symptoms***

Enter letter identifying the chapter of ICD-10 and up to 4 digits. Leave blank if none.

Specify: ....

2.079 ***Subclinical infection or nutritional problem***

If likely to be a contributory cause of any somatoform item in Section 2 that have been rated as present, rate 1.

Specify ....

SKIP => to 3.001 if optional checklist is not used.

OPTIONAL CHECKLIST (items 2.080 - 2.135)

List of somatoform symptoms

Rate this list only if items **2.010 - 2.035** are not sufficient for research purposes. The criteria for ratings these items are the same as those for items **2.010 - 2.035**. In general, physical disease already should have been excluded. Rate PS for past two years.

0 Symptom absent.
1 Symptom present. No full medical explanation.
2 Symptom present but probably due to physical disease.
8 Uncertain whether symptom is present.

Have you had a lot of trouble from any of the following symptoms?

2.080 ***Abdominal or stomach pain.***

2.081 ***Back pain, neckache.***

2.082 ***Pains in the joints.***

2.083 ***Pains in arms or legs.***

2.084 ***Chest pains or burning in the chest.***

2.085 ***Pressure in the chest.***

2.086 ***Headaches, tension in head or scalp, bursting feeling.***

2.087 ***Pain in any other part of the body.***

2.088 ***Pain moving from one place to another. Wandering pains.***

2.089 ***Vomiting or regurgitation.***

2.090 ***Vomiting throughout pregnancy.***

2.091 ***Nausea, indigestion, churning, fluttering, stomach ache.***

2.092 ***Bad taste in mouth, coated tongue.***

2.093 ***Lump in throat or difficulty swallowing.***

2.094 ***Loose bowels or diarrhoea.***

2.095 ***Constipation.***

2.096 ***Excessive gas or wind; bloated; belching; swollen.***

2.097 ***Several kinds of food cause R to feel ill.***

2.098 ***Shortness of breath when not making much effort.***

2.099 ***Heart pounding when not exercising; palpitations.***

2.100 ***Dizziness or light-headedness.***

2.101 ***Double vision.***

2.102 ***Blurred vision, not due to lack of proper glasses.***

2.103 ***Blindness in one or both eyes.***

2.104 ***Ringing or buzzing or noises in the ears/head, deafness.***

2.105 ***Loss of sensation in arm or leg.***

2.106 ***Loss of sensation elsewhere.***

2.107 ***Numbness or tingling sensations.***

2.108 ***Crawling or creeping sensations.***

2.109 ***Feelings of heaviness or lightness anywhere in the body.***

2.110 ***Difficulty walking.***

2.111 ***Periods of muscular weakness, e.g. problems lifting things.***

2.112 ***Paralysis, inability to move part of the body.***

2.113 ***Lost voice for 30 minutes or more.***

2.114 ***Other speech problems.***

2.115 ***Spells or fits of bodily shaking, lasting minutes.***

2.116 ***Fainting spells following dizziness.***

2.117 ***Spell of unconsciousness, trances, seizures, fits.***

2.118 ***Loss of memory for hours or more.***

2.119 ***Blotchiness of the skin, burning, itching.***

2.120 ***Frequency of passing water, burning urine.***

2.121 ***Pain or burning sensation when passing water.***

2.122 ***Difficulty passing water, or retention.***

2.123 ***Flushes or blushing.***

2.124 ***Pain on intercourse.***

2.125 ***Genital discharge.***

2.126 ***Genital discomfort/burning sensations, etc.***

2.127 ***Dry mouth.***

2.128 ***Excessive sweating.***

2.129 ***Irregular menstruation.***

2.130 ***Excessive menstrual bleeding.***

2.131 ***Excessive pain with menstruation.***

2.132 ***Impotence.***

2.133 ***Lack of sexual feeling.***

2.134 ***Feeling continually unwell/sickly/out of sorts.***

2.135 ***Any other bodily symptom, specify: ....***

2.136 ***Relation of somatoform symptoms to panic attacks***

0 Somatoform symptoms or panic attacks (**4.020**) absent.
1 Somatoform symptoms and panic attacks always occur together.
2 Somatoform symptoms occur at different times to panic attacks on at least some occasions.
3 Somatoform symptoms always occur at different times to panic attacks.
8 Uncertain.
9 NK.

If necessary, re-rate ages of onset of somatoform symptoms at items **2.071** and **2.072.**

3 Worrying, tension, etc.

Before rating any item that might be present, always establish whether there has been a period of 'normality' before onset. Symptom ratings require there to have been a definite deviation from a previous state when the symptom was absent.

Use Scale I to rate all Section 3 items except **3.004**.

First I should like to ask about some very common experiences and see whether you have had any of them recently.

3.001 ***Worrying***

Have you worried a great deal during [PERIOD]?

- *What is it like when you worry?*
- *Do unpleasant thoughts go round and round in your mind?*
- *Do you worry more than is necessary, given the problem?*
- *What happens when you try to turn your attention to something else?*
- *Can you stop worrying by looking at TV or reading or thinking about something you usually enjoy?*

A round of painful thought which cannot be stopped and is out of proportion to the topic of worry. Worries 'too much' but only in relation to real problems = mild.

3.002 ***Feeling of nervous tension***

Have you often felt on edge or keyed up or mentally strained?

- *What is that like?*
- *How severe is it?*
- *Do everyday problems get on top of you?*
- *Do you tend to startle too easily?*

There is no need for autonomic accompaniments for this symptom to be rated present though they usually are. Include exaggerated startle response. Feels strain only in relation to real problems = mild. If R says s/he is anxious but does not describe autonomic symptoms, consider rating here.

3.003 ***General muscular tension***

Have you had difficulty in relaxing during [PERIOD]?

- *Do your muscles feel tensed up?*
- *Where do you feel it?*
- *Can you relieve it by relaxing?*

Tension only in relation to real problems = mild.

If 3.001 - 3.003 are rated 0 ask 3.004.

3.004 ***Calmness in the face of problems***

These questions have been about nervous problems. Could I ask you the opposite kind of question? Would you say you were more calm and collected, less prone to irritability, restlessness, self-consciousness or nervous fatigue than most people during [PERIOD]?

- *That you would only get upset if something really serious happened to cause it?*

Probe for items **3.011** - **3.013** as appropriate.

1 Less prone to Section 3 problems than most.
2 About average.
3 More nervous than average.
4 Severe nervousness.

CUT OFF => to 4.001 if 3.001 - 3.003 rated 0 and 3.004 rated 1 or 2

3.005 ***Localized tension pains***

Have you had aches and pains, like headaches, neckache, backache, aching muscles, during the [PERIOD]?

- *What is it like?*

'Band round head', 'pressure', 'tightness in scalp', 'ache in back of neck' etc, etc not migraine or other specific syndrome. If probable physical cause, use etiology option.

3.006 ***Subjectively described restlessness***

Have you been so fidgety and restless that you couldn't sit still?

- *Did you have to keep pacing up and down?*

Fidgety = mild. Restlessness = moderate. Pacing = severe. Use judgement in rating. Do not automatically accept R's reply.

3.007 ***Fatiguability and exhaustion***

Have you been getting exhausted and worn out during the day, even when you haven't been working very hard?

Exclude hypersomnia = **8.016**. Unduly tired = mild. Feels worn out, effort to carry on = moderate. Completely exhausted = severe.

3.008 ***Sensitivity to noise***

Have you found that noise upsets you?

- *Do noises that other people find tolerable seem to penetrate or go through your head?*
- *More than ordinary dislike of loud noises?*
- *Do you have to put up with a lot of noise?*
- *Are you able to work or think in noisy places?*

Exclude ordinary dislike of loud noise. Sensitive but copes = mild.

3.009 ***Irritability***

Have you been very much more impatient or irritable than usual during [PERIOD]?

- *How has that shown itself?*
- *Do you keep it to yourself or raise your voice or flare up without reason?*
- *Have you really lost your temper or your control?*

1 Mild irritability, unusual for R, or brief domestic quarrels out of the ordinary.
2 Raised voice, anger, shouting, more frequently picking quarrels.
3 Pushing, hitting, lost control.

3.010 ***Simple ideas of reference***

Are you self-conscious in public?

- *Do you get the feeling that other people are taking particular notice of you, for example in the street or in a restaurant*
- *Are they really taking a special interest in you or is it perhaps that you are sensitive?*

R must recognize that the feelings originate from within but nevertheless cannot help thinking people talk, laugh, criticism, notice etc. Delusions of reference = 5.

3.011 ***Suspiciousness***

Have you tended to be more suspicious than usual?

See definition in Glossary.

3.012 ***Depersonalization and derealization***

Have you felt that things around you, or other people, or you yourself were unreal? [see **16.006 - 16.009**]

Exclude if solely in context of anxiety; see **4.026**. If any evidence for depersonalization, complete Section 16.

3.013 ***Non-delusional jealousy***

Have you been more inclined to be jealous? [check at **19.015**]

See Glossary definition.

3.014 ***Dates of PERIOD of Section 3 symptoms***

It is necessary to record only if dates of symptoms in this Section are different from PERIOD/S rated in Section 1 (items **1.007, 1.008, 1.009**).

If PERIOD used is RE or LB, record start and finish dates of PERIOD of symptoms in this Section. If PS (and dates of PE) used only record start date of PE:

Onset of PS/PE (mmyy)

Onset - offset of RE or LB to

3.015 ***Interference with activities due to Section 3 symptoms***

You have mentioned [summaries symptoms] *during PERIOD. Overall, how much interference has there been with your everyday activities because of these problems?*

Rate interference due to symptoms in Section 3.

0 No Section 3 symptoms present in significant degree.
1 Symptom/s present but little interference.
2 Moderate or intermittent interference.
3 Severe to incapacitating interference.

Opportunity to rate attribution of physical (including drug, substance etc) cause is present using the optional attributional scale and the dashed boxes; the user may also re-rate organic cause at the section/syndrome level at Section 13.

4 Panic, Anxiety and Phobias

If present but judged to be due to physical cause, e.g. hyperthyroidism, stimulants, withdrawal from drugs (e.g. Benzodiazepines), etc., users may rate individual items using the optional attribution scale or rate entire section at end of Section 4 or in Section 13. Consider also stress reaction in Section 13.

If R says s\he is anxious but does not describe autonomic symptoms, consider rating elsewhere, e.g., Section 3.

4.001 ***General rating of anxiety***

Now I should like to ask about feelings of anxiety or attacks of panic during PERIOD.

When people get anxious or panicky they often feel very fearful. They may feel their heart beating fast, or may they start shaking or sweating, or feel they can't get their breath. Have you had feelings like that? (Can you describe it?)

0 Anxiety and panic attacks absent.
1 Anxiety and/or panic attacks present.

4.002 ***General rating of phobias***

Some people have phobias. They feel anxious, or panicky or scared in certain situations, like being afraid of heights, or open spaces, or certain animals or insectsspiders, or in some social situations. They try to avoid them or even think about them.

Would that be true of your experience?

0 Phobias absent.
1 Phobias may be present.

How long have you noticed this? When did it first start?

Check ages at **4.056** and record duration of period at **4.059** unless same as in item **1.007**.

Symptoms of anxiety or panic are listed in items **4.003 - 4.019.** Those marked + are autonomic arousal symptoms specified in ICD-10 as especially important for panic states. The computer program sums them separately.

The list may be presented by the interviewer one item at a time, or as prompt card, to R directly. In either case, the terminology used should be appropriate for local circumstances and culture. For example, 'butterflies' will only be understood in a few parts of the world. Other examples should be found for the same symptom.

Some respondents may attest to these symptoms in addition to other somatic complaints but attribute them to a disabling physical cause for which they have sought medical help, or if that is not available have engaged in self treatment with physical remedies. Use somatoform checklists in Section 2 and rate 2. if panic attacks (**4.020**) are also rated present in this Section.

I should like to ask about a list of symptoms that some people get when they are anxious. Do you have any of these?

Rate items **4.003 - 4.019**.

0 Absent.
1 Symptoms present in isolation. Not in panic attack
2 All symptoms rated as occurring together during worst spell or attack. In Panic Attack.

4.003 *Can't get breath and smothering feeling.*

4.004+ *Heart pounding, missing beats, faster.*

4.005 *Dizzy, light-headed, faint, unsteady.*

4.006 *Tingling, numbness in face/fingers.*

4.007 *Tightness, discomfort or pain in chest.*

4.008+ *Dry mouth not due to medication or dehydration.*

4.009 *Difficulty in swallowing, lump in throat.*

4.010+ *Sweating, e.g. palms.*

4.011+ *Trembling or shaking, e.g. of hands or limbs.*

4.012 *Hot or cold sweats or flushes.*

4.013 *Unreality, 'not really here'. Like an actor.*

4.014 *Churning stomach, nausea, butterflies.*

4.015 *Fear of dying.*

4.016 *Feeling of choking.*

4.017 *Fear of going crazy, or fear of losing emotional control or passing out.*

4.018 *Apprehension, jumpiness, or increased startle response.*

4.019 *Other, e.g. urinary frequency, etc.*

CUT OFF => to 5.001 if 4.001 and 4.002 are rated 0, and no anxiety symptoms are present.

If no evidence of anxiety symptoms, ask:

You suggested that you did have problems with [anxiety and/or phobias] but you have not had any of these anxiety symptoms during PERIOD. Is that because you have not been in situations that provoke them?

SKIP => to 4.027 **if no anxiety symptoms occur during PERIOD because of avoidance.**

4.020 ***Frequency of panic attacks with autonomic symptoms***

Have you had any attacks of panic, or sudden attacks of anxiety, with unpleasant feelings like these (4.003 - 4.019), which very quickly became intolerable?

- *Can you describe a typical recent attack?*
- *How often did these attacks occur, say in a 4 week PERIOD?*
- *Did you only get them in certain situations?*
- *What situations?*

Include if they occur out of the blue, or in phobic situations, or apparently triggered by alarming thoughts or internal sensations. There is usually a sudden onset with a rapid crescendo to maximum. Consider the possible contribution of a cardiothoracic or other physical disease and if an attributable cause, consider using the etiology options. Nevertheless, it should be possible to decide on whether panic attacks have occurred, but if there is doubt about this symptom, rate 88.

Record approximate number in a 4 week period up to 40.

00 = No attacks.
40 = 40 or more.
55 = Psychotic symptoms make rating difficult.
88 = NK.

SKIP => to 4.023 **if no panic attacks with autonomic symptoms.**

4.021 ***Enduring apprehension of having another panic attack***

After an attack, did you worry about its effects or about getting another attack?

0 - 3 Use Scale I

4.022 ***Action to prevent or to end panic attack***

Did you have to do something to end the attack, like running out of the house, leaving a bus, phoning someone, calling a neighbor, opening the window to try to get air, etc.

0 Absent.
1 Action to prevent an attack.
2 Action to end an attack.
3 Action to prevent and to end an attack.

4.023 ***Free-floating anxiety***

Have you had anxiety feelings for longer periods, not just in attacks, during [PERIOD]?

- *Are there times when you are free of anxiety feelings?*
- *How often do you have them?*
- *How severe are they?*
- *Do they occur only in certain situations? Which?*

Exclude panic attacks and anxiety associated only with phobias. Include anxiety with no apparent reason that persists after the experience of panic attacks. Panic attacks, phobias and free floating anxiety can be present together within the same period. Duration is rated at **4.025**.

0 No free floating anxiety in PERIOD.
1 Occasional mild anxiety.
2 Moderately severe.
3 Severe autonomic anxiety.
5 Psychotic symptoms make rating difficult.

4.024 ***Anxious foreboding with autonomic symptoms***

Have you been anxious that something terrible might happen - like illness, death, accident or ruination - to you or to someone close to you?

- *Without any real cause for the feeling?*
- *Was that during PERIOD?*
- *Were the anxious or panicky feelings present?*

Include only if anxiety is the chief feature. If context is depressive rate at **6.022**.

0 Not present.
1 Excessive anxiety but only in real situations.
2 Moderately severe.
3 Severe.
5 Psychotic symptoms make rating difficult.

SKIP => to 4.026 if 4.023 and 4.024 are negative.

4.025 ***Duration of free floating anxiety***

How long have these feelings lasted? When did you first notice them?

Enter number of months up to 80.

81 = more than 80.
88 = NK.
99 = NA.

4.026 ***Depersonalization/derealization with anxiety***

Have you felt recently as though the world was unreal, or that you yourself were not real?

- *What was that like?*
- *Was it part of the anxiety? Did the anxiety change in any way when this feeling came on?*

Consider depersonalization in the context of anxiety or panic only. Sometimes the anxiety diminishes as depersonalization begins. This item is not the same as **3.012** or **16.006 - 16.009**.

0 - 3 Use Scale I

SKIP => to 4.059 if phobias definitely absent

PHOBIAS

[Continue from answer to **4.002**].

Could I ask you about phobias. Do situations or objects that would not bother most people make you anxious or panicky? Do you try to avoid these situations because they are so unpleasant and distressing?

Phobias are listed in items 4.027 - 4.045.

The list may be presented by the interviewer one item at a time, or as a prompt card, to the respondent directly. In either case, the terminology used should be appropriate for local circumstances and culture. For example, train phobias will rarely (but possibly) be found in areas with no railways. Use examples appropriate to the meaning of each item and to the circumstances and culture in which the interview takes place.

Rate only items that have caused distress and intense fear and/or anxiety symptoms at some time since onset of the disorder. Avoidance is very common but not essential for identification.

Phobias should only be rated 2 if associated with autonomic symptoms at some time since the disorder began, either in feared real or imagined situations (e.g., trains in a place with no railways).

R **must** acknowledge that their fear is excessive or unreasonable.

If associated with obsessional symptoms, rate phobias here and the nature of the association at **5.008**.

If associated with affective symptoms, rate at **6.022**.

Rate phobia items **4.027 - 4.045**.

0 Absent.
1 Present with Section **3** symptoms. Subclinical phobia.
2 Present with (tension) or anxiety symptoms.
severe.

SITUATIONAL PHOBIAS (usually when alone)

4.027 *In a public place when alone, open spaces, empty streets.*

4.028 *Crowds, shops, theaters, no easy exit.*

4.029 *Going out or travelling alone: buses, trains, planes.*

4.030 *Travelling away from home.*

4.031 *Collapsing while alone or no help near.*

4.032 *Being alone indoors.*

SOCIAL PHOBIAS (dread of behaving in an embarrassing or humiliating way when the focus of attention.)

4.033 *Eating or drinking or writing, etc., in public. Using public toilets.*

4.034 *The possibility of blushing, shaking, micturition, defecation or vomiting in public.*

4.035 *Speaking to or being in a small group of known people.*

4.036 *Speaking in public, or to strangers, or on phone.*

SPECIFIC (SIMPLE) PHOBIAS

4.037 *Storms, thunder, lightning.*

4.038 *Water - pool, lake, sea, baths, showers.*

4.039 *Animals: insects, feathers, birds, snakes, fur, rats, etc.*

4.040 *Enclosed spaces - lifts, tunnels, phone boxes, etc.*

4.041 *Flying.*

4.042 *Heights - bridges, balconies, windows in tall buildings, stairs.*

4.043 *Situations associated with illness: blood, injections, dental treatment, hospitals, death.*

4.044 *Fear of contracting illness, cancer, VD, AIDS, etc.*

4.045 *Other.* Specify............................

Exclude dysmorphophobia, rated at **16.011**, and hypochondriasis, rated at **2.038** and **2.039**.

SKIP => to 4.059 if no evidence for phobias.

4.046 ***Most severe situational phobia***

Which was the most severe of [situational phobias]?

- *How much distress or disability did it cause?*
- *How much of the time was it present?*

Use Scale I

4.047 ***Avoidance of most severe anxiety provoking situation***

Do (did) you try to avoid [situation]?

Use Scale I

4.048 ***Interference with activities due to most severe situational fear***

How much interference has there been with your everyday activities, work, or social relationships?

0 No interference.
1 Symptom/s present but little interference.
2 Moderate or intermittent interference.
3 Severe to incapacitating interference.

4.049 ***Most severe social phobia***

Use Scale I

4.050 ***Avoidance of most severe anxiety provoking social situation***

Do (did) you try to avoid [social situation]?

Use Scale I

4.051 ***Interference with activities due to most severe social fear***

How much interference has there been with your everyday activities, work, or social relationships?

0 No interference.
1 Symptom/s present but little interference.
2 Moderate or intermittent interference.
3 Severe to incapacitating interference.

4.052 ***Most severe specific phobia***

Use Scale I

4.053 ***Avoidance of most severe anxiety provoking specific stimulus***

Do (did) you try to avoid [situation]?

Use Scale I

4.054 ***Interference with activities due to most severe specific fear***

How much interference has there been with your everyday activities, work, or social relationships?

0 No interference.
1 Symptom/s present but little interference.
2 Moderate or intermittent interference.
3 Severe to incapacitating interference.

4.055 ***Relation of panics to phobias***

Did the panics you mentioned always occur with one of these phobias, or did you also have them without any warning?

0 No phobias (or no panics).
1 Only phobias.
2 Phobias with panic.
3 Panic rarely or never with phobias.

4.056 ***Age at first onset of panics or phobias***

How old were you when these panics/phobias first caused you real distress or limited your activities?

4.057 ***Phobias began after an episode of schizophrenia***

0 = No.
1 = Yes.

4.058 ***Phobias began after episode of major affective disorder***

0 = No.
1 = Yes.

If both anxiety and depression together, rate **6.022**.

If both anxiety and obsessional symptoms together, rate **5.008**.

4.059 ***Dates of PERIOD/s of Section 4 symptoms***

It is necessary to record only if dates of symptoms in this Section are different from PERIOD/S rated in Section 1 (items **1.007, 1.008, 1.009**). If different type of Section 4 disorders have occurred at different times rating of separate periods will provide more detailed information.

Onset of PS/PE (mmyy)

Onset - offset of RE or LB   to

4.060 ***Interference with activities due to Section 4 symptoms***

You have mentioned [summaries symptoms] *during PERIOD. Overall, how much interference has there been with your everyday activities because of these problems?*

Rate interference due to symptoms in Section 4.

0 No Section 4 symptoms present in significant degree.
1 Symptom/s present but little interference.
2 Moderate or intermittent interference.
3 Severe to incapacitating interference.

Although opportunity to rate attribution of physical (including drug, substance etc) cause is present using the optional attributional scale and the dashed boxes, items **4.061** and **4.062** allow the user to re-rate organic cause at the section / syndrome level. For example, consider effects of amphetamines, thyrotoxicosis and if present whether these provide a sufficient reason for the items rated.

4.061 ***Organic cause of anxiety symptoms in section***

Use the four criteria listed in Glossary.

0 Absent.
1 Probable organic cause; not fully confirmed.
2 Definite; confirmed by expert investigation.
8 Uncertain whether organic or not.

4.062 ***Identify organic cause of anxiety symptoms***

Following ICD-10 definitions, exclude alcohol and other substance disorders from rating here. If they are thought to cause Section 4 symptoms, rate in Section 11 or 12 or using individual item etiology option. For example, rate anxiety provoking effects of amphetamine etc., at **12.034**. At **4.062** enter letter identifying the chapter of ICD-10 and up to 3 digits. Leave blank if none.

5 Obsessional symptoms

The ideas and impulses are experienced as entering the mind against conscious resistance. R tries to resist them but fails. They are unpleasantly repetitive. They are acknowledged as excessive or unreasonable and recognized as part of R's own thoughts and much distress is caused by this, since the thoughts may be embarrassing or blasphemous. Thoughts (obsessions) and actions (compulsions) are closely connected and can be rated together under each item. In very chronic conditions, the resistance may lose force; therefore take history into account. Enter age of first onset at **5.013**.

1 **This rating is not appropriate in the case of obsessions and should not be used.**
2 Most days for at least 2 weeks, moderate severity.
3 As 2 but severe.
5 Difficult to differentiate from psychotic symptoms.

NB: Differentiate from autistic spectrum disorders (rate 0).

Ask these general probes:

Some people find that they have to keep on checking things that they know they have done; like gas taps, light switches, whether the front door is locked, and so on. Do you have problems like that?

Another problem with some people is that they have to keep things round them in a special order, far beyond ordinary tidiness. Is that true of you?

And what about keeping things clean? Do you have to spend a lot of time washing things repeatedly that are already clean?

- *There are other difficulties of a similar kind, like unpleasant and unwanted thoughts or images coming into the mind, which can't be resisted. Has that been a problem?*

If any evidence that there might be section 5 symptoms, probe further:

- *Can you describe what it is like?*
- *Where do these ideas or impulses come from?*
- *Do you try to resist them - for example, to stop yourself from ... ? (What happens?)*
- *Do you get very upset or distressed when you can't control it?*
- *How severe has it been?*
- *How often has it occurred during PERIOD? Most days? For as long as a fortnight? Longer?*

5.001 ***Evidence for obsessional and compulsive symptoms***

0 No evidence.
1 Sufficient to proceed below cut off.

CUT OFF => to 6.001 if no evidence for Section 5 symptoms.

Rate items **5.002 - 5.006** on Scale I, omitting the rating of 1.

When did the symptoms start? How old were you? Is this the first time you have had them?

Enter ages at **5.013**.

Use the probes to explore and rate the following symptoms:

5.002 ***Obsessional checking and repeating***

Checking gas taps, locks, light switches, etc. Touching or counting things many times. Repeating the same action over and over again.

5.003 ***Obsessional actions associated with excessive orderliness***

Keeping personal belongings neat and tidy far beyond the ordinary. Adopting routines to make sure that everything is in precisely the right place.

5.004 ***Obsessional thoughts of harm or accidents***

Unpleasant thoughts or words or images keep coming into the mind, against resistance. Thoughts of violence, or images of knives or fragments of glass.

5.005 ***Obsessional actions associated with cleanliness***

Washing repeatedly, against resistance. Distressed by thoughts of germs or other contamination.

5.006 ***Obsessional doubts, ruminations and feelings of incompletion***

Do not include non-specific indecision due to slow or muddled thinking, rated in **7.003**.

R's mind is full of doubts and questions about own actions and thoughts. Absurd thoughts keep intruding against resistance. Unpleasant feelings that tasks have not been fully completed. Constant questions about the meaning of everything, what even trivial events really mean. All symptoms rated must satisfy the general criteria for obsessions. Include obsessional preoccupation with ideas and images, blasphemous thoughts in church, etc., if there is a feeling of compulsion against resistance.

5.007 ***Fading of conscious resistance to obsessional symptoms***

1 No longer any conscious resistance.
2 One or more symptoms against resistance.

5.008 ***Relation of anxiety to obsessional symptoms***

If anxiety/phobic symptoms have been present at the same time as obsessional symptoms, which have been most clinically significant.

0 Anxiety symptoms absent or neither predominate.
1 Anxiety/phobic symptoms more clinically significant.
2 Obsessional symptoms more clinically significant.
8, 9

5.009 ***Relation of depressive to obsessional symptoms***

If symptoms of depression have been present at the same time as obsessional symptoms, which have been most clinically significant?

Rate here only if Section 6 has been taken first. Check **6.024**.

0 Absent, or symptoms of one type present only.
1 Depressive symptoms more severe or started first.
2 Obsessional symptoms more severe or started first.
8, 9

5.010 ***Relation of obsessions to compulsions***

If both obsessions and compulsions have been rated, decide which is more clinically significant:

0 No obsessions or compulsions rated, or only symptoms of one type rated.
1 Obsessions are more clinically significant.
2 Compulsions are more clinically significant.
8 Uncertain, or appear of equal importance.
9 NA.

5.011 ***Insight for obsessional symptoms***

0 No obsessions or ruminations rated.
1 R nearly always recognizes that obsessions and/or compulsions are unreasonable.
2 Sometimes recognizes that obsessions and/or compulsions are unreasonable.
3 For most of the episode, R does not recognize that obsessions and/or compulsions are unreasonable.

5.012 ***Content of obsessional symptoms limited to another disorder***

0 Content not limited to another disorder, or no symptoms.
1 Content of obsession is partially limited to another disorder.
2 Content is entirely limited, for example to depressive rumination or hypochondriasis.

5.013 ***Age at first onset of obsessional symptoms***

When were you first really distressed or handicapped because of these problems? How old were you then?

5.014 ***Dates of PERIOD/s of Section 5 symptoms***

It is necessary to record only if dates of symptoms in this Section are different from PERIOD rated in Section 1 (item **1.007, 1.008, 1.009, 1.010**).

Onset of PS/PE (mmyy)

Onset - offset of RE or LB to

5.015 ***Interference with activities due to Section 5 symptoms***

You have mentioned [summarise symptoms] *during PERIOD. Overall, how much interference has there been with your everyday activities because of these problems?*

Rate interference due to symptoms in Section 5.

0 No Section 5 symptoms present in significant degree.
1 Symptom/s present but little interference (up to 1 hour per day).
2 Moderate or intermittent interference (i.e. many hours per day).
3 Severe to incapacitating interference.

Although opportunity to rate attribution of physical (including drug, substance etc) cause is present using the optional attributional scale and the dashed boxes, items below allows the user to re-rate organic cause at the section / syndrome level.

5.016 ***Organic cause of obsessional symptoms in section***

Use the four criteria listed in Section 13.

0 Absent.
1 Probable organic cause; not fully confirmed.
2 Definite; confirmed by expert investigation.
8 Uncertain whether organic or not.

5.017 ***Identify organic cause of obsessional symptoms***

Following ICD-10 definitions, exclude alcohol and other substance disorders from rating here. If they are thought to cause Section 5 symptoms, rate in Section 11 or 12 or using individual item etiology option. At **5.017** enter letter identifying the chapter of ICD-10 and up to 3 digits. Leave blank if none.

6 Depressed mood and ideation

Remember that Sections do not always have to be taken in strict numerical order. If it is more appropriate to take items in Section 7 or 8 relevant to depressed mood first, that is the optimal order.

Symptoms of depression

Before rating any item that might be present, always establish whether there has been a period of 'normality' before onset. Symptom ratings require there to have been a definite deviation from a previous state when the symptom was absent.

Dysthymia

This is a condition of persistent despondency or gloom lasting for 2 years or more; with poor sleep, low energy, tedium vitae and a feeling of insufficiency. There can be periods of normal mood and basic coping is not impaired. If answers to the first questions suggest a chronic condition, complete Section 6, and also items **6.044 - 6.070**.

Recurrent Brief Depressive Disorder

Complete Section 6. Then rate items **6.072 - 6.076**.

May I ask some questions about feelings of sadness or depression?

6.001 ***Depressed mood***

Have you been feeling low in spirits recently/[PERIOD]?

- *Would you describe your mood as sad, downcast, gloomy, despairing or deeply depressed?*
- *Has it been mild, moderate or severe?*
- *How much of the time were you in reasonable spirits?*
- *And how much of the time were you really low?*
- *When did you last feel your usual self?*

When rating clinical severity of depression, remember that deeply depressed people may not necessarily cry, and that retardation may produce an impression of apathy and empty mood (see **6.007**). Irritability (item **3.009**) may also mask depression. Rate mood on subjective description of peak period. Manifestations at interview are rated at **23.002**. Consult Glossary.

Use Scale I

SKIP => to 6.003 if 6.001 < 2.

6.002 ***Duration of Depressed mood***

How long have you been feeling like this?
When was the last time you usually felt free of this feeling?

Enter number of weeks up to 80 (88 = unsure, but always try to rate approximate duration conservatively, i.e. if it is at least a month, rate 04 and not 88):

If 6.001 is rated between 1 - 3, ask:

What do you think is the cause?

Note the answers for items in Section **13**.

If > 2 years, consider dysthymia.

6.003 ***'Masked' depression***

If **6.001** is rated 0 or 1, but the interviewer considers that a depressed mood is masked because of other symptoms (e.g. irritability), or difficulty in introspecting (e.g. cognitive impairment from any cause, or educational problem), or a cultural tendency to manifest depression in other forms than mood, e.g. in somatic symptoms rate here. If uncertain, rate 8. See Glossary.

1 Circumstances suggest that depressed mood might be present but a definite rating cannot be made [continue beyond cut-off].
2 Masked depression of mood very probably present.

6.004 ***Tearfulness and crying***

Do you often feel like crying?

- *Do you actually cry?*
- *How often does that happen?*
- *For how long has it been going on?*
- *What starts you off?*

Mild	=	tears in eyes;
Moderate	=	tearfulness;
Severe	=	long periods of weeping, nearly every day, much of the day. Being unable to cry often suggests severity, but is not included here.

Use Scale I

If > 2 years, consider dysthymia.

6.005 ***Capacity for enjoyment***

Have you been able to positively enjoy things like taking a walk, working at your hobbies or interests, having a nice meal with friends, winning a game or receiving a compliment?

Or do you seem unable to find pleasure even in things you used to enjoy?

- *How much of the time were you unable to enjoy things as much as usual?*
- *When did you last really enjoy something? What?*
- *Do you keep up the appearance of enjoyment?*

This should be a definite loss compared with the normal state but the loss need not have begun during PERIOD.

1 Positively enjoys life.
2 Little or no loss, but not 1.
3 Marked loss.
4 Total loss.

If > 2 years, consider dysthymia

CUT OFF => to 7.001 if no evidence of depressed mood, or of other Section 6 symptoms.

If Section 6 symptoms may be present without a context of depression, complete the relevant items.

If depressed for more than two years, continue but consider dysthymia **6.044 - 6.071** throughout the section unless clearly severe and episodic.

6.006 ***Loss of hope for the future***

How do you see the future?

- *Do you think there are some good days still to come?*
- *How much of the time does everything seem quite hopeless?*
- *When did you last feel you could see a reasonable future? What were you looking forward to?*

This should be a definite loss compared with the normal state but the loss need not have begun during PERIOD.

Use Scale I

Consider dysthymia if > 2 years.

6.007 ***Feeling of loss of feeling***

Sometimes people don't describe sadness or depression as such but say they have lost the ability to feel any emotion at all. They can't feel sad and can't cry. Have you actually experienced that lack?

- *What is it like compared with your ordinary mood?*
- *How severe or continuous was it during the PERIOD?*
- *Have you been free of it at all?*

This should be a definite loss compared with the normal state but the loss need not have begun during PERIOD. 'The loss of feeling is felt, the numbness perceived, the lifelessness experienced' (E Bleuler). See Glossary definition. Constricted affect but not as a positive feeling = 1. There is no equivalent trait.

Use Scale I

6.008 ***Unremitting depression***

Does the depressed mood come and go, or does it seem to be always there, even if only in the background?

- *Does it get better or worse with circumstances or is it always much the same?*

0 Absent.
1 Reactive to circumstances on most days.
2 Variable and/or but not event-reactive.
3 Persistent through most of period; not reactive.

6.009 ***Morning depression***

What time of day does the depression feel worst?

- *Is there any time when it is less severe?*

0 No depression, or not worst early.
1 Regularly feels worst early in the day.

6.010 ***Preoccupation with death or catastrophe***

Do you tend to brood over possible disasters, like death or ruin or some catastrophe that could occur to you or others?

- *How much of the time have you been free of this?*
- *How often was the brooding really intense and continuous?*

If delusional, rate at 6.019.

Use Scale I

If > 2 years consider dysthymia

6.011 ***Suicide or self-harm***

Have you thought about harming yourself or even made an attempt at suicide, during [PERIOD]?

- *What happened?*

Include whether thought due to depression or not:

0 Absent.
1 Deliberately considered suicide or self-injury (intrusive thoughts) but made no attempt.
2 Injured self or made an attempt but no serious harm resulted.
3 Serious self-harm (not part of suicide attempt).
4 Made an attempt at suicide designed to result in death.

If 6.011 is rated 0 also, ask 6.012.

6.012 ***Tedium vitae***

Have you felt, during the [PERIOD], that life was not worth living or that you would not care if you didn't wake in the morning?

- *Would you even wish to have some fatal disease or accident?*
- *How much of the time did the feeling last?*
- *Recurrent thoughts of death, death-wish?*

Use Scale I

6.013 ***Pathological guilt***

Do you tend to blame yourself for something you have done or thought; to feel guilty or ashamed of yourself?

- *What is it that you think you have done wrong?*
- *How much of the time, in [PERIOD], have you been free of it?*
- *How often did you feel guilty?*

Rate only if guilt is not realistic; not if actions have been blameworthy and guilt proportional. **6.013** and **6.014** can co-exist. If so, rate both.

If delusional, rate at 6.018.

Use Scale I

6.014 ***Guilty ideas of reference***

Do you have the feeling that you are being blamed or accused by others because of some action or lapse or deficiency that you yourself feel was blameworthy?

- *How much of the time, in the [PERIOD] have you been free of the feeling?*
- *How often have you had the feeling that you were being blamed for something really serious?*

Rate only if guilt is not realistic. Omit if actions have been blameworthy and reference could be just. If delusional, rate at **19.010**.

Use Scale I

SELF ATTITUDES

6.015 ***Loss of self-confidence with other people***

How confident do you feel in yourself? For example, in talking to others or in managing your relationships with other people?

- *When could you last feel confident in that way?*
- *Was there any change or has it always been like that?*

This should be a definite loss compared with the normal state but it need not have begun during Period.

Use Scale I

If 6.015 is trait like consider rating of 27.033.

If > 2 years consider dysthymia.

6.016 ***Social withdrawal***

Have you wanted to stay away from other people? Why? Have you been suspicious of their intentions? Of actual harm?

Check item **3.011.**

Use Scale I

If 6.016 is prolonged or life-long consider rating of 27.034.

6.017 ***Loss of self-esteem***

What is your opinion of yourself compared to other people?

- *Do you seem to feel less competent than they are?*
- *In what way?*
- *Do you feel inferior, or even worthless?*
- *Is that new or have you always felt that way?*

This should be a definite change compared with the normal state but the change need not have begun during PERIOD. Worthless = 3.

Use Scale I

If 6.017 is prolonged or life-long consider rating of 27.033.

If > 2 years consider dysthymia.

PSYCHOTIC AFFECTIVE SYMPTOMS

Part II of PSE10, from which the following 4 items are taken, should always be completed if any psychotic symptom is present.

6.018 ***Delusions of guilt or worthlessness in context of depression***

Unshakeable conviction, in context of depression, of guilt, crime, evil, harm to others, worthlessness, etc., for which R feels culpable and deserving of punishment.

1 Rare 2 Occasional 3 Frequent [**19.025**].

6.019 ***Delusions of catastrophe in context of depression***

Unshakeable conviction, in context of depression, that the world is about to end, the country is evil and will be destroyed, that R or family will be ruined, etc. R feels responsible.

1 Rare.
2 Occasional.
3 Frequent [**19.026**].

6.020 ***Hypochondriacal delusions in context of depression***

Unshakeable conviction that bowels are stopped up, insides are rotting, etc. in context of depressed mood.

1 Rare.
2 Occasional.
3 Frequent [**19.027**].

6.021 ***Congruence of auditory hallucinations with affective state*** [Check at **17.010**]

Rate congruence if, for example, content based on guilt: 'You deserve to die you sinner' etc., or on delusions of grandeur: 'Go to the palace. You are the King', etc. If unsure, rate 8. Include if content was congruent earlier but mood has improved before AH disappeared.

0 No auditory hallucinations (AH).
1 Virtually all content congruent with affective state.
2 More congruent than not.
3 Congruent and incongruent AH equally.
4 More incongruent than congruent.
5 Virtually all content incongruent.

GENERAL RATINGS OF DEPRESSION

To rate the following items it is not necessary to make a firm diagnosis of depression. The items should be rated if there is any possibility of the diagnosis. The computer program will make the decision and a clinical diagnosis can be entered at Section 7 of CHS (**27.071 - 27.082**).

If both anxiety and depression present:

6.022 ***Depression or anxiety primary***

You have mentioned that you have felt both anxious and depressed [in PERIOD]. You mentioned ... [summarise symptoms for R]

- *Some people say they are depressed because they have phobias or anxiety; they are miserable because of that. If the phobias or anxiety cleared up they would not be depressed any more.*
- *Other people say that if only the depression would go, they would stop being anxious.*
- *Can you decide if one or the other is more important for you or is there little to choose between them?*

Probe further. Use examples below if appropriate.

Anxiety and depression must both be rated definitely present for this item to be meaningful.

1 Anxiety is primary. Depression appears to be entirely explicable in terms of the limitations placed on the subject by the symptoms of anxiety, e.g. being unable to leave the house, travel, meet people, etc., or being afraid of heart disease because of palpitations.
2 Anxiety and depression both present but seem independent of each other, or fluctuating in predominance; e.g. onset of depression makes any preceding anxiety more severe, or vice versa.
3 Depression is primary. Anxiety is either a result of the depression (e.g. subject is frightened because of morbid or suicidal ideas) or it takes the form of fears of catastrophe, forebodings about illness or death, dread of having to face the day when first waking in the morning, preoccupation that something awful is going to happen. Panic attacks and situational anxiety, if present, are secondary to depression.
8 Uncertain, e.g. because of lack of information.

Reconsider judgement of hierarchical relationship of symptoms in Section 13.

6.023 ***Relation of somatoform to depressive symptoms***

Respondents with somatoform symptoms may deny depressive symptoms. Information from informatmants and records may be required. Review, if necessary, ratings of **6.003** and **6.005**. If symptoms of depression have been present at the same time as somatoform symptoms, which were most severe or started first?

0 Depressive symptoms more severe or started first.
1 Somatoform symptoms more severe or started first.
8 Uncertain.
9 Symptoms not occurring together.

6.024 ***Relation of obsessional to depressive symptoms***

If symptoms of depression have been present at the same time as obsessional symptoms, which were most severe or started first? [Check **5.009**].

0 Depressive symptoms more severe or started first.
1 Obsessional symptoms more severe or started first.
8 Uncertain.
9 Symptoms not occurring together.

6.025 ***Age of first onset of depressive symptoms***

How old were you when these feelings of depression first started?

When did you first feel really distressed by them or they first started to limit your daily activities?

Check dates of LB and first onset at **1.011** and **1.015** and duration current depressive symptoms PE at **6.026**. If PE represents a relapse/recurrence, record dates of most recent episodes at **6.038 - 6.043**.

6.026 ***Dates of PERIOD/s of Section 5 depressive symptoms***

It is necessary to record only if dates of symptoms in this Section are different from PERIOD rated in Section 1 (item **1.007, 1.009, 1.010**).

Onset of PS/PE (mmyy)

Onset - offset of RE or LB to

6.027 ***Interference due to depression [exclude trait ratings]***

How much interference has there been with your everyday activities because of depression?

- *What sort of problem is it?*

If an in-patient, consider period up to admission. Check at **13.006**.

0 No significant depression present.
1 Symptoms present but little interference.
2 Moderate or intermittent interference.
3 Severe to incapacitating.

6.028 ***Organic cause of depressive symptoms***

Use the four criteria listed in Section 13.

0 Absent.
1 Probable organic cause; not fully confirmed.
2 Definite; confirmed by expert investigation.
8 Uncertain whether organic or not.

6.029 ***Identify organic cause of depressive symptoms***

Following ICD-10 definitions, exclude alcohol and other substance disorders for rating here. If they are thought to cause depressive symptoms, rate in Section 11 or 12 or using optional attribution scale. At **6.029** enter ICD-10 chapter identifier and up to 3 digits. Leave blank if none.

SKIP => to 6.044 if R now in first ever episode or if never depressed.

HISTORY ITEMS

It is essential to rate **6.030 - 6.038**, if there is any possibility that a depressive syndrome is present. For recurrent episodes (i.e. Seasonal, rapid cycling) rate **6.037** and **6.072 - 6.077**. The actual classification will be made by the computer program.

Rate the whole course if information is available.

6.030 ***Episodes of major affective disorder***

There have been at least two episodes of major affective (manic, mixed and/or depressive) disorder during the course (including PS), demarcated by a switch to an episode of opposite or mixed polarity, or separated by a period of normal mood of at least 2 months.

0 Not rated.
1 Only one affective episode in whole course.
2 At least 2 affective episodes.
3 3 + episodes.
4 4 + episodes within a 12-month period.
5 Most recent 4 episodes within a 12-month period.
8 Uncertain.
9 Not Applicable. No affective episode during course.

6.031 ***Personality prior to depression onset***

0 Within normal range.
1 Significant personality disturbance before first major depressive episode.

SKIP => to 6.034 if 6.030 is rated 1.

If 6.030 is rated 2-5 (At least 2 affective episodes):

6.032 ***Severity of affective episodes***

How severe have episodes been, in general?

0 Not rated.
1 Mild episodes predominant.
2 Moderate episodes predominant.
3 Severe episodes predominant.
8 Uncertain which were predominant.
9 No episodes/ not applicable.

If 6.030 is rated 2 (2 or more affective episodes):

6.033 ***Two or more depressive episodes with recovery***

Have there been at least two major depressive episodes, each followed by recovery for 2 months or more? Consider the whole clinical course.

0 Not rated.
1 Only one such episode.
2 At least 2 such episodes.
3 Complete remission occurred between 2 most recent episodes.
8 Uncertain.
9 NA.

If 6.033 is rated 2 (At least 2 such episodes) rate 6.034.

6.034 ***Response to adequate antidepressive therapy***

Have symptoms in the Major Depressive Disorders shown a good response to adequate antidepressive therapy?

0 Not rated.
1 Response has not been satisfactory.
2 Response has been satisfactory.
8 Uncertain.
9 No episodes.

6.035 ***One or more manic or mixed episodes during the course***

Has there been at least one manic, hypomanic or mixed episode during the course (including PS/PE)?

0 Not rated.
1 There have been no hypo/manic or mixed episodes.
2 At least one manic or hypomanic or mixed episode.
8 Uncertain.
9 NA.

6.036 ***Mixed episodes during the course***

Rate if mixed episodes present currently or in the past (see Glossary for definition of mixed episode).

0 Not rated.
1 There have been no mixed episodes.
2 One or more mixed episodes in the past.
3 Current episode mixed.
8 Uncertain.
9 NA.

6.037 ***Dates of 6 most recent episodes of depressive disorders before PE***

This rating is required for classification of seasonal, rapid cycling and other recurrent episodes involving clinically significant symptoms from Sections 6, 7, 8.

0 Depressive Episode list not required and not used to date episodes of affective disorder.
1 Dates of up to 6 (most) recent depressive episodes prior to PE recorded or PS recorded
2 As 1, but separate SCAN ratings made of previous episode/s.
8 Uncertain: respondent unable/unwilling to give satisfactory account of past episodes of depression.
9 Not applicable: reasons........................

6.038 ***Previous depressive episode 1 dates*** to

6.039 ***Previous depressive episode 2 dates*** to

6.040 ***Previous depressive episode 3 dates*** to

6.041 ***Previous depressive episode 4 dates*** [ ][ ][ ][ ] to [ ][ ][ ][ ]

6.042 ***Previous depressive episode 5 dates*** [ ][ ][ ][ ] to [ ][ ][ ][ ]

6.043 ***Previous depressive episode 6 dates*** [ ][ ][ ][ ] to [ ][ ][ ][ ]

CHECKLISTS (items 6.044 - 6.077)

PERSISTENT DEPRESSIVE STATES, DYSTHYMIA

Dysthymia consists of persistent despondency or gloom lasting for at least 2 years; with poor sleep, low energy, tedium vitae, and brooding. However, there is basic coping (though a feeling of insufficiency), and occasional brief periods of normal mood. Most items in the checklist have been covered in Section 6 or elsewhere in the PSE, and the interviewer will have noted which have occurred frequently over a two year period.

It will be useful to consult case records and/or an informant.

Rate only for period of 2 years before interview. Do not record dates in optional depressive episode lists **6.038 - 6.043, 1.019 - 1.024.**

[**6.044** and at least three of the features **6.045 - 6.061** should be present during at least some of the episodes during the two years.]

Rate items **6.044 - 6.061**:

0 No.
1 Yes, but isolated occurrence. (Checklist items covered elsewhere in the PSE fulfil the designated severity criteria)
2 Present at same time as all other items rated 2 (**6.044 - 6.061**).
9 NK.

6.044 ***2+ years depression; remissions for few weeks only*** [ ]

(0 = no, 1 = yes, 9 = uncertain).

Skip => to 6.072 if 6.044 = 0

6.045 ***A reduction in energy or activity*** (**7.006**, see **22.003, 22.005**). [ ]

6.046 ***Insomnia*** (**8.009** and **8.011 - 8.016**). [ ]

6.047 ***Hypersomnia*** **(8.016)** ☐

6.048 ***Loss of self-confidence, feelings of inadequacy or low self-esteem*** **(6.015, 6.017).** ☐

6.049 ***Difficulty in concentrating or making decisions*** **(7.002, 7.003).** ☐

6.050 ***Frequent tearfulness, often in tears*** **(6.004).** ☐

6.051 ***Generalised loss of interest or enjoyment in sex and other in pleasurable activities*** **(6.005).** ☐

6.052 ***Loss of interest or enjoyment in sex*** **(8.024** and **8.025)** ☐

6.053 ***Feeling of hopelessness or despair*** **(6.006).** ☐

6.054 ***Decreased productivity, effectiveness, or perceived inability to cope with the routine responsibilities of everyday life*** **(7.007).** ☐

6.055 ***Pessimistic about the future or brooding over the past*** **(6.006** or **6.010).** ☐

6.056 ***Social withdrawal*** **(6.016).** ☐

6.057 ***Less talkative than normal. Reduced talkativeness*** (see **24.016**). ☐

6.058 ***Poor appetite or over-eating*** **(8.005; 8.008**, Section 9). ☐

6.059 ***Chronic fatigue or tiredness*** **(3.007)** ☐

6.060 ***Feelings of guilt, brooding about the past*** **(6.013)** ☐

6.061 ***Subjective feelings of irritability or excessive anger*** **(3.009)** ☐

6.062 ***Interference with activities due to symptoms rated at 6.044 - 6.061*** ☐

You have mentioned [summarise symptoms] *during PERIOD. Overall, how much interference has there been with your everyday activities because of these problems?*

Rate interference due to symptoms in Section 6.

0 No Section 6 symptoms present in significant degree.
1 Symptom/s present but little interference.
2 Moderate or intermittent interference.
3 Severe to incapacitating interference.

6.063 ***Age when present episode of dysthymia began.***

6.064 ***Did an episode of major depression precede the present episode of dysthymia?***

6.065 ***If 6.064 = 1, did a period of full remission, lasting 6 months or more, intervene between the two episodes?***

6.066 ***Did a major depressive syndrome begin during the first 2 years of the present episode of dysthymia? Rate even if unsure whether full criteria for major depressive episode have been met.***

6.067 ***Age at first onset of any episode of dysthymia.***

6.068 ***Did an episode of major depression precede the first episode of dysthymia?***

6.069 ***If 6.068 = 1, did a period of full remission, lasting 6 months or more, intervene between the previous two episodes?***

Rate contribution of physical cause (including drugs, substance etc) and/or psychosocial cause of dysthymia symptoms using etiology option for individual items or below for group.

6.070 ***Organic cause of dysthymia symptoms***

Use the four criteria listed in Section 13.

0 Absent.
1 Probable organic cause; not fully confirmed.
2 Definite; confirmed by expert investigation.
8 Not known whether organic or not.

6.071 ***Identify organic cause of dysthymia symptoms***

Following ICD-10 definitions, exclude alcohol and other substance disorders from rating here. If they are thought to cause a brief depressive phase, rate in Section 11 or 12. At **6.071** enter letter identifying the chapter of ICD-10 and up to 3 digits. Leave blank if none.

Consider cyclothymia at 10.031 - 10.057.

RECURRENT BRIEF DEPRESSIVE DISORDER

Rate the following items.

0 Absent.
1 Present.
8 NK.

6.072 ***There has been a brief depressive phase at least once a month during the past year.***

6.073 ***Such phases of depression have lasted less than two weeks (usually only 2 - 3 days) during past year.***

6.074 ***Such phases have not occurred solely in association with the menstrual cycle during last week of the luteal phase and remitted within a few days of the onset of menses.***

6.075 ***Interference with activities due to brief recurrent depressive disorder.***

0 Syndrome absent or not rated.
1 Syndrome present, but little difference.
2 Moderate or intermittent interference.
3 Severe to incapacitating interference.

Rate attribution of physical cause (including drugs, substance etc) below for group.

6.076 ***Organic cause of brief depressive phase***

Use the four criteria listed in Section 13.

0 Absent.
1 Probable organic cause; not fully confirmed.
2 Definite; confirmed by expert investigation.
8 Uncertain whether organic or not.

6.077 ***Identify organic cause of brief depressive phase***

Following ICD-10 definitions, exclude alcohol and other substance disorders from rating here. If they are thought to cause a brief depressive phase, rate in Section 11 or 12. At **6.077** enter letter identifying the chapter of ICD-10 and up to 3 digits. Leave blank if none.

7 Thinking, concentration, energy, interests

Before rating any item that might be present, always establish whether there has been a period of 'normality' before onset. Symptom ratings require there to have been a definite deviation from a previous state when the symptom was absent. A trait rating means that the condition has 'always' been present; i.e. from childhood or early adolescence.

[Consider mild cognitive impairment or decline. If evidence, use Section 21].

Now may I ask you about changes in your energy, interests and concentration?

- *Have you been able to think as clearly as usual during [PERIOD]?*
- *Can you concentrate as usual - for example read the newspaper, or watch a TV program through, or attend to one of your ordinary activities?*
- *What was your concentration like at your best?*
- *What about energy and interests?*

If any problem:

When did it start or have you always been like it?

7.001 ***Positive cognitive functioning***

Perhaps it is quite the opposite? You have plenty of energy, lots of interests and very good concentration? Or would you say you were about average during the [PERIOD]?

1 Feels less prone to Section 7 problems than average. Plenty of interests, concentration, energy.
2 Little or no loss, but not 1.
3 Marked loss.
4 Almost total loss.

CUT OFF => to 8.001 if no evidence of Section 7 symptoms in PERIOD

7.002 ***Loss of concentration***

Has your concentration been as good as usual during the [PERIOD] or does your attention wander?

- *Do you find you can't read an article in the paper or watch a TV program right through even though you could previously have done so?*
- *Do you find that you cannot concentrate sufficiently to complete any task properly; e.g. cooking, conversation, work?*
- *How often has it been really severe?*

There should be a definite loss compared with the normal state but it need not have begun during the PERIOD.

1 Mild. Has to concentrate much harder than usual but can then complete task fairly satisfactorily.
2 Moderate or intermittently severe impairment.
3 Severe impairment much of the time.

If >2 years consider dysthymia.

7.003 ***Subjectively inefficient thinking***

Have you been able to think quite clearly in [PERIOD]?

Can you make up your mind as well as usual or do simple decisions seem to have become very difficult?

- *Have you found that your thoughts are much slower?*
- *Have you been more muddled or indecisive?*
- *Could you think much more clearly before?*
- *How often has the difficulty been really severe, so that you could hardly begin to think out any problems?*

There should be a definite loss compared with the normal state but it need not have begun during the [PERIOD].

1 Mild. Enough to be noticeable but can, with an effort, think clearly enough.
2 Moderate or intermittently severe impairment.
3 Severe impairment much of the time.

Recall at **14.003** and **18.002**.

7.004 ***Loss of interests***

What were your interests before?

- *Have you lost interest in your work, or your recreations?*
- *Have you lost interest in your dress and appearance?*
- *How much of the time were you free of these problems?*
- *How often was it really bad, so that you lost your interests altogether?*
- *When could you last take your normal interest in things?*

This should be a definite loss compared with the normal state but it need not have begun during the [PERIOD].

1 Mild. Enough to be noticeable but does, with an effort, maintain fairly normal interests.
2 Moderate or intermittently severe impairment.
3 Severe impairment much of the time.

If > 2 years consider dysthymia.

7.005 ***Subjective feeling of retardation***

Have you felt as though you were slowed down in your movements, as though everyone and everything else was moving much faster?

- *Have you seemed to talk or move more slowly than usual?*
- *Have your arms or legs felt heavy? (like lead?)*
- *How much of the time in the [PERIOD] were you free of this?*
- *How often has it been really bad so that you hardly seemed able to move at all?*

There should have been a definite onset compared to normal but it need not have begun during [PERIOD].

1 Mild. Enough to be noticeable but can with an effort put the feeling more or less out of mind.
2 Moderate or intermittently severe impairment.
3 Severe impairment much of the time.

If > 2 years consider dysthymia.

7.006 ***Loss of energy (drive)***

Have you felt that you have lost energy or vigor?

- *Have you lost a vital spark, as though everything was too much trouble; that you couldn't bother?*
- *How long have you had this during the [PERIOD]?*
- *How energetic were you when you last felt normal?*

There should be a definite loss compared with the normal state but the loss need not have begun during the [PERIOD].

1 Mild. Some loss of energy but can manage to appear usual self when the situation requires it.
2 Moderate or intermittently severe impairment.
3 Severe impairment much of the time.

If > 2 years consider dysthymia.

7.007 ***Feeling of being overwhelmed by everyday tasks***

Does it seem as though every little task, no matter how trivial, is just too much responsibility to bear?

- *Do you feel that you just can't cope?*
- *That it is painful even to think about having to start the next task of the day?*
- *Can you describe that?*
- *When did you last feel normal in this respect?*

There should be a definite loss compared with the normal state but the loss need not have begun during the period.

1 Mild. Enough to be noticeable but can, with an effort put the feeling out of mind.
2 Moderate or intermittently severe impairment.
3 Severe impairment much of the time.

If > 2 years consider dysthymia.

7.008 ***Dates of PERIOD of Section 7 symptoms*** to

It is necessary to record only if dates of symptoms in this Section are different from PERIOD rated in Section 1 (items **1.007, 1.009, 1.010 - 1.013**).

7.009 ***Interference with activities due to Section 7 symptoms***

You have mentioned [summarise symptoms] *during PERIOD. Overall, how much interference has there been with your everyday activities because of these problems?*

Rate interference due to symptoms in Section 7.

0 No Section 7 symptoms present in significant degree.
1 Symptom/s present but little interference.
2 Moderate or intermittent interference.
3 Severe to incapacitating interference.

Although opportunity to rate attribution of physical (including drug, substance etc) cause is present using the optional attributional scale and the dashed boxes, items below allow the user to re-rate organic cause at the section/syndrome level.

7.010 ***Organic cause of Section 7 symptoms***

Use the four criteria listed in Section 13.

0 Absent.
1 Probable organic cause; not fully confirmed.
2 Definite; confirmed by expert investigation.
8 Uncertain whether organic or not.

7.011 ***Identify organic cause of Section 7 symptoms***

Following ICD-10 definitions, exclude alcohol and other substance disorders from rating here. If they are thought to cause a brief depressive phase, rate in Section 11 or 12 or using individual item etiology option. At **7.011** enter letter identifying the chapter of ICD-10 and up to 3 digits. Leave blank if none.

8 Bodily functions

Use information from Section 2 if relevant. If the item is present, and meets criteria, but it was the examiner`s judgement that it was probably caused by physical illness or medication, rate presence/severity in this Section and attribution of cause in Section 13.

8.001 ***Weight in kilograms***

What is your present weight? [1 kg = 2.2 lb].

8.002 ***Height in cms***

What is your height now? [2.5 cms = 1 in].

8.003 ***Weight increase or decrease since adolescence (kgs)***

What was your weight when you were a young person, say before you were 20?

0 No.
1 Yes.

8.004 ***Record estimated weight age 18 to 20 (Kgs) and record weight now at 8.001. Weight increase since adolescence (kgs)***

8.005 ***Change in appetite***

What is your appetite like?

- *How long has it been poor or increased?*
- *What was it due to?*
- *Has it been associated with recent symptoms?*

1 Marked loss of appetite for at least 2 weeks.
2 Marked loss of appetite 2-4 weeks.
3 Marked loss of appetite for more than a month.
4 Increased appetite, less than 2 weeks.
5 Increased appetite, more than 2 weeks.
6 Appetite always poor.
7 Fluctuation in appetite (mixed 1 - 5).

8.006 ***Loss of weight [from usual] in period***

Has there been any change in weight during [PERIOD]?

- *Did you lose or gain?*
- *What was the most you gained or lost in a month?*

0 No change.
1 Moderate loss (not 2).
2 More than 5% of body weight in a month or equivalent (i.e. 3.5+ kg if 70 kg).
3 More than 15% body weight loss.
4 Lack of weight gain expected to occur [e.g. in growing child or adolescent, or after recovery from severe weight loss due to physical factors] of equivalent severity to 3.

8.007 ***Gain of weight [from usual] in period***

There can be both gain and loss of weight in one episode of affective disorder.

0 No change.
1 Moderate gain (not 2).
2 More than 5% of body weight in a month or equivalent (i.e. 3.5+ kg if 70 kg).

8.008 ***Weight problem*** *[Note for Section 9 on eating disorders]*

Have you had a serious problem with weight?

- *Apart from times when you were physically ill or depressed?*
- *What sort of problem?*

0 No.
1 Yes.

SLEEP PROBLEMS

Have you had any difficulty or disturbance in sleeping during [PERIOD] or have you been sleeping too much?

- *Has the quality of your sleep been as good as usual?*

Use judgement if night or shift work.

- *Do you take sleeping tablets?*

If yes:

Why? What are they? How much do you take? For how long? Who prescribes them? Do you sleep well while taking them?

CUT OFF => to 8.024 **if no sleep problems, not taking tablets, and not depressed during period.**

If depressed mood:

8.009 ***Sleep problem with depressed mood***

Have the sleep problems been associated mainly with the time you have been depressed?

0 Not specially associated.
1 Definite association with depressed mood.

If > 2 years consider dysthymia.

8.010 ***Effect of tablets on sleep pattern***

0 Not taking sleeping tablets/sedatives.
1 Sleeps well with tablets.
2 Sleeps poorly in spite of tablets.

Ask the questions at **8.011**. There is no need to ask all items **8.012 - 8.019** if the description in response to **8.011** is clear. If > 2 years consider dysthymia.

8.011 ***Delayed sleep***

What is your usual pattern of sleeping? [When to bed, when wake?]

- *How long ago was it normal for you?*
- *What is/was the problem now/during PERIOD?*

Allow R to answer fully then choose the appropriate items. Do not ask questions R has already answered spontaneously.

- *Is the problem one of falling asleep or of quality?*
- *How long do you lie awake? As much as an hour?*
- *How often does that happen? As much as 3 times a week?*
- *Is the sleep problem distressing?*
- *Has it been going on for as long as a month?*
- *Does it reduce efficiency during the day?*

0 No problem.
1 R complains of delay (at least an hour), not 2.
2 Delay of 1 hour+, 3+ times a week, at least 1 month.

If sleep problem judged to be due to physical condition, e.g. pain, psychoactive substance use disorder, medication, neurological disorder etc., rate attribution of cause using etiology options.

If there is poor quality sleep ask **8.012**.

8.012 ***Poor quality sleep***

In what way is your sleep unsatisfactory?

- *Do you wake feeling sleepy even when you have not lain awake or woken frequently?*

Do not include reasons for disturbed sleep such as nightmares, etc., rated elsewhere.

0 No problem. Good quality sleep.
1 Sleep is not refreshing (specify in what way).
2 3+ times a week, for at least a month.

8.013 ***Middle insomnia***

Do you wake during the night?

- *For how many hours? How often? For what period of time?*

0 No problem.
1 R complains of waking (at least an hour) but not to criterion (2).
2 Waking for 1 hour+, 3+ times a week, for 1+ month at a time.

8.014 ***Early waking***

What time do you usually wake in the morning when you are sleeping normally?

- *Have you been waking much earlier than this?*

Use frequency and time probes, making due allowance for unusual working hours.

If R has been depressed during period:

Does/did it happen only when you are depressed?

0 No problem.
1 R complains of early waking (at least an hour) but not to criterion (2).
2 Wakes at least 2 hours early, 3+ times a week, for at least a month.
3 As (2), but only in association with episode of depression.

8.015 ***Disturbance of normal sleep-wake cycle***

Have you found that your usual sleep pattern is reversed, that you want to sleep when you should be awake and are wakeful when you should be sleeping? [Frequency and time probes]

0 No problem.
1 R complains of symptoms but not to criterion (2).
2 Nearly every day for at least a month.

8.016 *Hypersomnia*

Do you find that you are very sleepy during the daytime and that you have attacks of sleep that you can't resist? [When R would normally have been awake].

Or that there is a long period of sleepiness or confusion after you wake?

- *Does that happen more or less every day?*
- *Has it lasted as long as a month?*
- *Does it happen only because you are not sleeping at night?*

Exclude daytime sleeping if due only to lack of sleep at night.

0 No problem.
1 Excessive sleep but not to (2).
2 More or less daily for at least a month.

If sleep problem attributed to physical condition; consider e.g. narcolepsy, (see Glossary), rate cause in Section 13 or using optional attributional scale.

8.017 *Disturbing dreams or nightmares*

Have you had any disturbing dreams or nightmares?

- *Do you wake in panic or with a cry? [Frequency probes]*

0 No problem.
1 R complains of nightmares but not to (2).
2 R is distressed by sleep disturbance resulting from frequent nightmares.

8.018 *Sleep terrors*

R gets up from sleep with a panicky scream and intense anxiety manifested in a autonomic arousal. May rush out of bed screaming. Limited recall of event. May be disoriented for a few minutes and at risk of injury. Usually first third of sleep. Perseverative movements and unresponsive for a time even to familiar others. Less than 10 minutes. [Frequency and time probes].

0 No problem.
1 Sleep terrors repeatedly present (2 or more) during PERIOD.

8.019 ***Sleepwalking***

Has anyone said you walked in your sleep [during PERIOD]?

- *What did they say happened?*
- *Did you remember anything when you woke?*

Usually first third of sleep period. Blank face, unresponsive, difficult to wake, amnesia after waking, only brief confusion on waking and no impairment following.

0 None.
1 Repeated episodes during period.

8.020 ***Dates of PERIOD/S of Section 8 sleep symptoms***

to

Record only if dates of sleep symptoms in this Section are different from PERIOD rated in Section 1 (items **1.007, 1.009, 1.010 - 1.013**).

8.021 ***Interference with activities due to Section 8 sleep symptoms (items 8.009 - 8.019)***

You have mentioned [summarise symptoms] *during PERIOD. Overall, how much interference has there been with your everyday activities because of these problems?*

Rate interference due to sleep symptoms in Section 8.

0 No Section symptoms present in significant degree.
1 Symptom/s present but little interference.
2 Moderate or intermittent interference.
3 Severe to incapacitating interference.

Rate attribution of physical (including drug, substance etc.) cause using attributional scale options or below for group.

8.022 ***Organic cause of Section 8 sleep problems***

Use the four criteria listed in Section 13.

0 Absent.
1 Probable organic cause; not fully confirmed.
2 Definite; confirmed by expert investigation.
8 Uncertain whether organic or not.

8.023 ***Identify organic cause of sleep problems***

Following ICD-10 definitions, exclude alcohol and other substance disorders from rating here. If they are thought to cause sleep problems, rate in Section 11 or 12 or using optional attributional scale for individual items. At **8.023** enter letter identifying the chapter of ICD-10 and up to 3 digits. Leave blank if none.

8.024 ***Sexual dysfunction***

Has your interest in sex been a lot less than usual during [PERIOD]?

- *Has your activity been much reduced?*
- *If you think about sex do you feel desire?*
- *Do you have sexual fantasies as much as before?*

Rate sexual drive, not orientation.

1 Somewhat lower.
2 Markedly lower.
3 Virtually absent during [PERIOD].
7 Has never been actively interested.

If depressed ask **8.025**.

8.025 ***Loss of libido associated with depression***

Has this loss of interest mainly been associated with your mood?

0 Not specially associated.
1 Definite association with depressed mood.

8.026 ***Intercourse unpleasant***

Do you find that intercourse gives no pleasure, or is painful, or that you can't perform satisfactorily?

- *Any other problem - e.g. gender or sexual preference?*
- *Is [...] distressing for you? Would you like to change?*

0 No further sexual problem mentioned.
1 R does have other sexual problems.

If other sexual dysfunction problems suspected to be present, apply ICD-10 DCR criteria for F 52 and record in Clinical History Schedule optional checklist.

8.027 ***Dates of PERIOD/S of Section 8 sexual dysfunction symptoms*** ☐☐☐☐ to ☐☐☐☐

Record only if dates of psychosexual dysfunction symptoms are different from PERIOD rated in Section 1 (items **1.007, 1.009, 1.010 - 1.013**).

8.028 ***Interference with activities due to Section 8 sexual dysfunction symptoms*** ☐

You have mentioned [summarise symptoms] *during PERIOD. Overall, how much interference has there been with your everyday activities because of these problems?*

Rate interference due to sexual dysfunction symptoms in Section 8. Include disorders coded in Clinical History Schedule from ICD-10 F 52.

0 No sexual dysfunction symptoms present in significant degree.
1 Symptom/s present but little interference.
2 Moderate or intermittent interference.
3 Severe to incapacitating interference.

Rate attribution of physical cause (including drugs, substance etc.) and/or psychosocial cause of Section 7 symptoms using etiology option for individual items or below for group.

8.029 ***Organic cause of Section 8 sexual dysfunction*** ☐

Use the four criteria listed in Section 13.

0 Absent.
1 Probable organic cause; not fully confirmed.
2 Definite; confirmed by expert investigation.
8 Uncertain whether organic or not.

8.030 ***Identify organic cause of Section 8 sexual dysfunction*** ☐☐☐☐

Following ICD-10 definitions, exclude alcohol and other substance disorders from rating here. If they are thought to cause a brief depressive phase, rate in Section 11 or 12 or using individual item etiology option. At **8.030** enter letter identifying the chapter of ICD-10 and up to 3 digits. Leave blank if none.

9 Eating disorders

All those who have had weight or appetite problems due to dieting, or a preoccupation with body shape, should be asked Section 9 unless due only to physical disorder. The technical definition of 'underweight' is 15% below weight for age and height. However, the first section should be completed if there is any possibility of anorexia nervosa during past year; weight at key date may be misleading.

The period rated is usually the **past year**. A previous episode may also be rated if needed.

9.001 ***Dread of being or becoming fat***

Have you worried, during the past year, about eating too much, or putting on weight, or getting too fat?

- *Have you felt guilty about eating?*
- *Or been afraid of losing control of the amount you eat?*
- *Do you pay a lot of attention to your shape/figure?*
- *How much does this distress you?*

Include preoccupation with weight, shape, or quantity of food.

0 No.
1 Yes.

9.002 ***Earlier episode of undereating***

0 Never.
1 Has had earlier episode(s) of dread of becoming fat, with undereating.

9.003 ***Irresistible and persistent craving for food***

Have you had a feeling of craving for food in the past year or being unable to stop once started?

- *Even when you knew you were overweight?*
- *How persistent was it?*
- *Could you always resist it?*

0 No.
1 Yes.

9.004 ***Earlier episode of overeating***

0 Never.
1 Has had earlier episode(s) of craving for food and overeating.

If any eating disorder during past year ask 9.005:

9.005 ***Interference due to appetite problems***

How much interference has there been with your everyday activities because of these problems you have described?

- *When did they start? How old were you?*

0 No symptoms present in significant degree.
1 Symptoms present but little interference.
2 Moderate or intermittent interference.
3 Severe interference with everyday functioning.

Rerate this item if necessary after further interview.

CUT OFF => to 10.001 if no evidence for anorexia or bulimia.

ANOREXIA

9.006 ***Undue preoccupation with bodily shape***

Do other people think you are too fat?

- *Do they think you are too thin?*
- *Have you been very preoccupied with the feeling that you were too fat even though, strictly-speaking, you were underweight?*

Intensive over-valued idea persists. Self perception of being too fat. Preoccupied with body shape.

0 No.
1 Yes.

9.007 ***Imposes low weight threshold***

Refusal to maintain body weight at or above a minimally normal range (i.e. 85% of expected body weight).

0 Absent.
1 Present.

9.008 ***Avoidance of fattening foods***

What action have you taken to avoid getting too fat?

- *Have you avoided fattening foods?*
- *Have you any special dietary rules?*
- *Even though your weight is normal or low?*

0 No.
1 Yes.

9.009 ***Actions to lose weight through self restriction***

Have you tried to lose weight by exercising a great deal? Or by taking appetite suppressants/diet pills?

0 No.
1 Yes.

9.010 ***Actions to lose weight through purgation***

Have you tried to lose weight by making yourself sick, taking laxatives, or taking diuretics?

0 No.
1 Yes.

9.011 ***Prepubertal onset of anorexia***

Women: Presence of primary amenorrhea and/or delayed or arrested growth.
Men: Genitals remain juvenile and/or delayed or arrested growth.

0 Not present.
1 Present.

SKIP => to 9.013 if prepubertal onset.

9.012 ***Change in Sexual Function***

Women:

Have you missed any periods while you were keeping your weight down?

- *How many?*

Vaginal bleeding in anorexic women on replacement hormone therapy or the contraceptive pill is rated 8.

0 No.
1 Yes but not 2.
2 Absent menstruation for Loss of three or more consecutive periods: amenorrhea.
8 Uncertain.
9 Postmenopausal.

Men:

Have you noticed any change in sexual function while you were keeping your weight down?

- *Were you less interested or less able to perform?*

0 No.
1 Occasionally.
2 Frequently.

How long has this trouble been going on? When did it first start?

Enter ages at **9.017 - 9.020.** Code duration of present episode at **9.021**.

BULIMIA

9.013 ***Binge eating with a sense of loss of control***

Have you had periods when you would eat abnormally large amounts of food within an hour or so, that is, binge eating?

- *How many times a week, at its height?*
- *How long did that go on for?*
- *Do/did you feel out of control?*

0 None.
1 Binges but not 2.
2 Binges and feels out of control, but not 3 or 4.
3 Twice a week on average over a period of 3 months but not 4.
4 Binges as 3 but feels out of control.

If 9.003 and/or 9.013 is positive, ask: 9.014.

9.014 ***Dread of getting too fat in spite of craving***
[The same item as **9.006** in a different context].

Have you had a fear of getting too fat, in spite of this craving for food or loss of control once you start?

0 No.
1 Yes.

9.015 ***Actions to correct binging by purgation***

After these binges have you tried making yourself sick or taking laxatives or taking appetite suppressants?

0 No.
1 Yes.

9.016 ***Restrictive actions to correct binging***

After these binges have you tried strict dieting, fasting, starvation or excessive exercise?

0 No.
1 Yes.

When did these problems start?

Enter ages at **9.017 - 9.020**. Record duration of present episode at **9.022**.

HISTORY ITEMS

Review any positive items - **9.001, 9.003 => 9.006 - 9.016**.

When did these present problems begin?

- *When was the first time you became concerned about these problems? How old were you then?*
- *Was that because of dieting or ...*
- *And how did the problem develop?*
- *So you were about ... years old?*

9.017 ***Age present episode of anorexia started.*** ☐☐

9.018 ***Age present episode of bulimia started.*** ☐☐

9.019 ***Age at first onset of anorexia.*** ☐☐

9.020 ***Age at first onset of bulimia.*** ☐☐

9.021 ***Duration present episode of anorexia*** ☐☐☐☐ to ☐☐☐☐

9.022 ***Duration present episode of bulimia*** ☐☐☐☐ to ☐☐☐☐

9.023 ***Relation of bulimia to anorexia*** ☐

0 Bulimia absent or present only during anorexia.
1 Bulimia present at other times.

CUT OFF => to 10.001 if a previous episode is not rated.

If earlier episode(s) is rated, give dates:

9.024 ***Anorexia*** ☐☐☐☐ to ☐☐☐☐

9.025 ***Bulimia*** ☐☐☐☐ to ☐☐☐☐

Skip back => to 9.006 **and complete Section for previous episode.**

Skip back => to 1.001 **if Sections 1 - 8 have not been completed, and take Sections 10 - 14 subsequently.**

9.026 ***Interference with activities due to Section 9 symptoms***

You have mentioned [summarise symptoms] *during PERIOD. Overall, how much interference has there been with your everyday activities because of these problems?*

Rate interference due to symptoms in Section 9.

0 No Section 9 symptoms present in significant degree.
1 Symptom/s present but little interference.
2 Moderate or intermittent interference.
3 Severe to incapacitating interference.

Rate attribution of physical cause (including drugs,medical condition, etc) of Section 9 symptoms using attribution of etiology option for individual items or below for group.

9.027 ***Organic cause of Section 9 symptoms***

Use the four criteria listed in Section 13.

0 Absent.
1 Probable organic cause; not fully confirmed.
2 Definite; confirmed by expert investigation.
8 Uncertain whether organic or not.

9.028 ***Identify organic cause of Section 9 symptoms***

Following ICD-10 definitions, exclude alcohol and other substance disorders from rating here. If they are thought to cause a brief depressive phase, rate in Section 11 or 12 or using individual item etiology option. At **9.028** enter letter identifying the chapter of ICD-10 and up to 3 digits. Leave blank if none.

10 Expansive mood and ideation

If present but due to amphetamine or other stimulants, rate attribution of cause at **12.034:1 - 9**; may also rate using optional attribution of etiology scale and dashed boxes for individual items, and for group at Section 13.

Cyclothymia is a condition of 2 years or more of instability of mood, with remissions of less than 2 months, involving periods of depression and hypomania, not severe enough for major depression or mania during that time. Rate hypomanic element at **10.031 - 10.050**. Dysthymia is rated at **6.044 - 6.061**.

10.001 Expansive (elevated) mood

Have you sometimes felt intensely happy or elated, without reason, during [PERIOD]? So elated that it was unnatural?

- *Was it out of character for you?*
- *How long did it last? Days? More than a week?*
- *How much of the time was your mood normal?*
- *Have you been taking pep pills to make you 'high'?*

1 Mild.
2 Moderate.
3 Severe.

Use SCALE I.

10.002 Irritable mood

Have you sometimes found you were easily irritated, that any little problem provoked you, or that other people said you were much too impatient? Did you have a hasty temper or actually become aggressive for [PERIOD]?

Check irritability at **3.009**.

Rate irritability at **3.009**, suspiciousness at **3.011**.

1 Mild.
2 Moderate.
3 Severe.

Use SCALE I.

10.003 ***Duration of expansive or irritable mood***

1 Present, but less than four days.
2 Present at least 4 days consecutively.
3 Present, at least 1 week.
4 Present more than 2 months.
5 Hospital admission required irrespective of duration.

If > 2 years, consider cyclothymia [See note at top of Section].

CUT OFF => to 11.001 if 10.001, 10.002, 10.003, 3.009 and 3.011 all rated 0.

10.004 ***Pressing and racing thoughts***

Do you find thoughts crowding into and racing through your mind - as though they were speeded up and you had too many thoughts compared with usual - during [PERIOD]?

- *Could you describe that?*
- *How long did it last?*

Use SCALE I

10.005 ***Overtalkativeness***

Have people said that you talked too fast and too much so that they couldn't understand you? Or do you feel pressure to keep talking?

Use SCALE I

If > 2 years, consider cyclothymia.

10.006 ***Distractibility***

Have you found that you could not keep your attention on one subject long enough to deal with it properly?

- *That your attention keeps being distracted by irrelevant things happening around you?*

Use SCALE I

If > 2 years, consider cyclothymia.

10.007 ***Self-reported overactivity***

Have you been more active than usual - so active that you or other people thought something was wrong?

Use SCALE I

If > 2 years, consider cyclothymia.

10.008 ***Sharpened thinking***

Have you been able to think more clearly or crisply than usual in [PERIOD]? Have you been especially efficient in solving problems and thinking things through? (Have you been unusually creative?)

Ability to concentrate, solve problems, create, etc. seems unequivocally well beyond R's ordinary mode, but R does not necessary construe this increased mental facility as connoting superior or extraordinary intelligence when compared to others. This symptom is often accompanied by inflated self-esteem, but exaggerated self-esteem and grandiosity are rated separately in item **10.010**.

Use SCALE I

If > 2 years, consider cyclothymia.

10.009 ***Overly entertaining talk***

Have you become unusually witty in communicating with others? (Quick with verbal "comebacks"?)

- *Have you, indeed, been much funnier than is your custom?*
- *Have others found you to be extremely entertaining? (Have others found it difficult to take your jokes?)*

Persistent witty remarks or retorts, or inappropriate punning and joking behavior that often occur in the context of, but should be rated separately from, pressured talk (**10.005**).

Use SCALE I

10.010 ***Exaggerated self-esteem***

Have you seemed specially efficient at work or in your daily activities, as though you had super powers or talents?

- *How do you explain this?*

Feelings of health, competence, intelligence, talent, etc., beyond the ordinary but definable as exaggerations of abilities rather than as delusions, which are rated at **10.016 and 10.017**.

Use SCALE I

If > 2 years, consider cyclothymia.

10.011 ***Over optimism***

Of all the things you have done in your life, which ones are your best (most important) achievements? (Which others?)

- *How does the future look to you? Do you see great things coming?*
- *Have you ever had times when it seemed as if nothing could go wrong? (When you were incredibly lucky?)*

Unduly optimistic view of the past and future. Include retrospective falsification and exaggeration of past achievements that are neither facetious "tall tales" nor delusional convictions (item **10.011**). Also include a sense of infallible luckiness <u>not</u> based on R's perceived capabilities.

Use SCALE I

If > 2 years consider cyclothymia.

10.012 ***Actions based on expansive mood***

Have you spent a lot of money unwisely in PERIOD/ month? Or acquired property, or bought a lot of things you didn't need, or made many business investments?

- *Have any problems arisen? Do some people think you have been unwise?*
- *What did you do?*

Reckless or foolhardy actions in context of expansive mood, with no insight at the time. Sexual indiscretions are rated separately at **10.015**.

1 Moderate overspending, embarrassing social contacts, unwise promises, etc.
2 At least one serious problem.
3 Several serious problems e.g. reckless driving, buying sprees, leaving secure job, because of expansive mood.

If > 2 years consider cyclothymia.

10.013 ***Decreased need for sleep***

Use SCALE I

Have you been able to manage with far less sleep than usual during [PERIOD] without seeming to get tired?

- What is that like?

1	Mild:	one hour less.
2	Moderate:	2-3 hours less.
3	Severe:	>3 hours less sleep.

Use SCALE I

10.014 ***Socially embarrassing behavior***

Have you been more sociable than usual?

- *In what way?*

Include undue or inappropriate sociability, over-familiarity, socially interfering, or loss of social inhibition.

Use SCALE I

If > 2 years, consider cyclothymia.

10.015 ***Increased sexual drive or activity.***

Have you found your sex interest or activities changed much during [PERIOD]?

- In what way?

Use SCALE I

If > 2 years, consider cyclothymia.

PSYCHOTIC AFFECTIVE SYMPTOMS

Part two of PSE10, from which the following 2 items are taken, should always be completed if any psychotic symptom is present in the context of an expansive mood. Congruence of auditory hallucinations is rated at **6.021**. Rerate if necessary and compare with item **17.010**.

10.016 ***Delusions of grandiose abilities*** [in context of expansive mood] [check **19.029**]

1 Rare.
2 Occasional.
3 Frequent.

10.017 ***Delusions of grandiose identity*** [in context of expansive mood] [check **19.030**]

1 Rare.
2 Occasional.
3 Frequently.

10.018 ***Age at first onset of manic symptom*** [check **1.016**]

10.019 ***Mixed episode during the course*** [Check **6.036**]

Rate if mixed episodes present currently or in the past. (See Glossary for definition of mixed episode.)

0 Not rated.
1 There have been no mixed episodes.
2 One or more mixed episodes in past.
3 Current episode mixed.
8 Uncertain.
9 NA.

10.020 ***Dates of PERIOD/S of Section 10 symptoms*** □□□□ to □□□□

Record only if dates of symptoms in the Section are different from PERIOD rated in Section 1 (items **1.007, 1.009, 1.010 - 1.013).**

10.021 ***Dates of 6 most recent episodes of expansive mood and ideation before PE*** □

This rating required for classification of seasonal, rapid cycling and other recurrent episodes involving clinically significant symptoms from Sections 6, 7, 8. Do not include dates of PERIOD recorded in Section 10 (record at **10.020**).

0 Expansive mood episode list not required and not used to date episodes of affective disorder.
1 Dates of up to 6 (most) recent expansive mood episodes prior to PE or PS recorded.
2 As in 1, but separate scan ratings made of previous episode/s.
8 Uncertain: respondent unable/ unwilling to give satisfactory account of past episodes of elevated mood.
9 Not applicable: reasons......

Record dates of up to 6 most recent previous episodes, recording month and year of onset and recovery. These dates are required for classification based on course.

10.022 ***Previous episode 1 of expansive mood; dates*** □□□□ to □□□□

10.023 ***Previous episode 2 of expansive mood; dates*** □□□□ to □□□□

10.024 ***Previous episode 3 of expansive mood; dates*** □□□□ to □□□□

10.025 ***Previous episode 4 of expansive mood; dates*** □□□□ to □□□□

10.026 ***Previous episode 5 of expansive mood; dates*** □□□□ to □□□□

10.027 ***Previous episode 6 of expansive mood; dates*** □□□□ to □□□□

You have mentioned [summarise symptoms] *during PERIOD. Overall, how much interference has there been with your everyday activities because of these problems?*

Rate interference due to symptoms in Section 10.

0 No Section 10 symptoms present in significant degree.
1 Symptom/s present but little interference.
2 Moderate or intermittent interference.
3 Severe to incapacitating interference.

Rate attribution of physical cause (including drugs, medical condition, etc) of Section 10 symptoms using attributional etiology scale for individual items or below for group.

10.029 ***Organic cause of Section 10 symptoms*** ☐

0 Absent.
1 Probable organic cause; not fully confirmed.
2 Definite; confirmed by expert investigation.
8 Uncertain whether organic or not.

10.030 ***Identify organic cause of Section 10 symptoms*** ☐☐☐☐

Following ICD-10 definitions, exclude alcohol and other substance disorders from rating here. If they are thought to cause Seçtion 10 symptoms, rate in Section 11 or 12 or using individual item etiology option. At **10.030** enter letter identifying the chapter of ICD-10 and up to 3 digits. Leave blank if none.

Reconsider **6.030 - 6.043** if ratings of course of affective disorders need to be changed.

Rate interference with everyday activities at **13.011**.

CHECKLIST (items 10.031 - 10.057)

HYPOMANIC ASPECTS OF CYCLOTHYMIA. PERSISTENT MINOR AFFECTIVE DISORDER

Cyclothymia is rated on the basis of at least 2 years instability of mood, involving several periods of both depression and hypomania, with or without intervening periods of normal mood. Remissions should not last more than 2 months. Hypomania is rated below. Equivalent depressive symptoms have been rated at **6.044 - 6.061**.

It will be useful to consult case records and/or an informant.

Rate only for 2 years before interview.

If necessary, ask:

Have you had periods when you felt depressed and others when you felt unusually happy or 'high' when there was no reason for it? Not just ordinary high spirits in company or having a good time.

- *Do people say that your mood goes up and down unpredictably?*
- *How long have you had changeable moods like that?*
- *How long have you been free of symptoms during the past two years?*
- *When was your mood last quite stable?*
- *Have you had only periods when you were high, without any periods when you were low?*

DURING PERIODS OF PERSISTENT HYPOMANIA

Rate the following items:

0 No.
1 Yes.
8 NK.

10.031 ***Periods of elation*** **(10.001)**. ☐

10.032 ***Increased goal-directed activity*** **(10.007)**. ☐

10.033 ***Distractibility (attention easily drawn to irrelevant stimuli)*** **(10.006)**. ☐

10.034 ***Decreased need for sleep*** **(10.013)**. ☐

10.035 ***Inflated self-esteem*** **(10.010)**. ☐

10.036 ***Sharpened or unusually creative thinking.*** **(10.008).**

10.037 ***Increased sociability*** **(10.014).**

10.038 ***More talkative*** **(10.005).**

10.039 ***Thoughts racing*** **(10.004).**

10.040 ***More witty than usual*** **(10.009).**

10.041 ***Increased involvement in sex*** **(10.015).**

10.042 ***Increase in other pleasurable activities*** **(10.012).**

10.043 ***Over-optimism or exaggeration of past-achievements*** **(10.011).**

10.044 ***Periods of irritability*** **(10.002).**

10.045 ***More than 2 years of hypomania, remissions < 2 months.***

10.046 ***Age at onset of present cyclothymia.***

10.047 ***Age at first onset of cyclothymia.***

10.048 ***Age at first onset of hypomanic syndrome.***

10.049 ***2 + years of periods of depression*** **(6.001).**

10.050 ***2 + years of periods of instability of mood.***

Periods of depression (**10.049**) occur during the 2 year periods of elation or irritability rated at items **10.001** and **10.002**. Symptom-free periods last less than 2 months.

CUT OFF => to 11.001 if 10.050 is rated 0.

10.051 ***Age at onset of present cyclothymia.***

10.052 ***Did a major depressive or manic syndrome begin contiguous with and after full establishment of the present episode of cyclothymia?***

Rate even if unsure whether full criteria for a major depressive or manic episode have been met.

10.053 ***Age at first onset of cyclothymia.***

10.054 ***Did a previous major depressive or manic syndrome begin contiguous with and after full establishment of the previous episode of cyclothymia?***

Rate even if unsure whether full criteria for a major depressive or manic episode have been met.

10.055 ***Interference with activities due to persistent hypomania and of cyclothymia***

0 Absent.
1 Little interference.
2 Moderate or intermittent.
3 Severe to incapacitating.

Rate contribution of physical cause (including drugs, substance etc) to hypomania and cyclothymia using attribution of etiology scale option for individual items or below for group.

10.056 ***Organic cause of hypomania and cyclothymia symptoms***

0 Absent.
1 Probable organic cause; not fully confirmed.
2 Definite; confirmed by expert investigation.
8 Uncertain whether organic or not.

10.057 ***Identify organic cause of hypomania and cyclothymia symptoms***

Following ICD-10 definitions, exclude alcohol and other substance disorders from rating here. If they are thought to cause a brief depressive phase, rate in Section 11 or 12 or using individual item etiology option. At **10.057** enter letter identifying the chapter of ICD-10 and up to 3 digits. Leave blank if none.

11 Use of alcohol

Most items in Section 11 can be rated for two periods, the past year in the left hand box and 'lifetime before' in the right hand box. The latter ratings are mainly necessary when there have been no problems with drinking during the past year but there have been previously.

11.001 ***Lifetime abstention***

Do you drink alcohol at all?

If no:

Were you always an abstainer or did you drink in the past?

- *You have never taken alcohol? Not even a glass of wine with a meal or to celebrate a special occasion?*

1 Has never used alcohol.
2 Once or twice in a lifetime.

CUT OFF => to 12.001 if 11.001 is rated 1 or 2.

3 Alcohol used more than 2.

Continue:

First obtain a general account of the effects of alcohol, and whether any problems have been associated with drinking during the past year or formerly. The following probes should be used only to help respondent give a spontaneous account. Use local terms for drinks and drinking.

About how often do/did you take a drink? What sort of effects does/did alcohol have on you? Would it be difficult to do without it? (In what way?) Have you ever felt you had problems with alcohol?

- *Could you describe them?*
- *Did they affect your activities or your health? How?*
- *Was there a particular period when the problems were worse?*
- *When was that?*

This description may already answer some of the questions **11.002 - 11.013**. Ask only those that are necessary to complete the picture. **11.006 - 11.032** are concerned only with the past year; the rest should be rated for the past year and/or 'lifetime before', as appropriate.

11.002 ***Frequency of drinking, past year (PY) and lifetime before (LB)***

Quantity is not taken into consideration in this rating; e.g. the quantity taken on any 1 day could vary from 1 glass to 2 litres. Probe further and rate the most appropriate estimate from the list.

About how often did you take a drink during the past year?

0	Not at all during the period year.	5	3 - 4 days a week.
1	Less than once a month.	6	Nearly every day.
2	Once a month.	7	Every day.
3	2 - 3 days a month.	8	NK.
4	1 - 2 days a week.	9	NA.

11.003 ***Estimated usual daily grams of alcohol, past year (PY and LB)***

If necessary, use chart to convert local glass and bottle sizes. The standard drink contains about 12 grams of alcohol. Estimate for the usual intake, even if drinking is sometimes heavier.

What sort of alcohol do you usually drink?

- *On a day when you do drink, about how many glasses [pints of beer, glasses of wine, bottles, etc.] do you usually take?*
- *What size glass/bottle would that be?*
- *Would that be much the same on each occasion (day)?*

If drinks beer, ask:

Do you prefer strong special beer/lager?

1	1 - 2 Standard drinks.	6	20 - 25 Standard drinks.
2	3 - 4 Standard drinks.	7	25+ Standard drinks.
3	5 - 9 Standard drinks.	8	Not known.
4	10 - 14 Standard drinks.	9	Not applicable.
5	15 - 19 Standard drinks.		

11.004 ***Heavier daily alcohol, grams (PY and LB)***

Are/were there days when you drank much more than that? Like parties, celebrations, weekends or holidays? When was that?
How much did you drink on those days of heavier drinking?

Enter "heavier" daily intake during past year, and LB if heavier still, in grams, using the method of **11.003**.

1 1 - 2 standard drinks.
2 3 - 4 standard drinks.
3 5 - 9 standard drinks.
4 10 - 14 standard drinks.
5 15 - 19 standard drinks.
6 20 - 25 standard drinks.
7 25+ standard drinks.
8 NK.
9 NA.

11.005 ***Frequency of heavier daily alcohol (PY and LB)***

How often do/did you drink that heavier amount?

- *How often in the past year?*
- *How often during your previous heaviest period of drinking?*

Enter frequency of heavier daily intake during past year and during 'lifetime before', if greater.

1 Occasionally; less than once a month.
2 About once a month.
3 More than once a month but not 4.
4 Once a week or more.

11.006 ***Most recent drink***

When was the last time that you had a drink?

- *Even one of any kind?*
- *Today, yesterday?* (probe times)

1 Within 24 hours.
2 1 - 3 days ago.
3 4 - 7 days ago.
4 2 weeks to 1 month ago.
5 1 month to 6 months ago.
6 6 months to 1 year.
7 More than 1 year ago.
8 Uncertain.
9 Unknown.

Be particularly careful about asking questions **11.007 - 11.013**, if intake of alcohol has always been low and problems have been denied. There is no need to ask again if the questions have already been answered. But if there is any doubt, always probe fully.

11.007 ***Subjective need for alcohol (PY and LB)***

Have you ever felt that you needed alcohol and could not manage without it?

- *How strong is/was the feeling?*
- *Were you uneasy at times or in places when alcohol was not available?*
- *In situation where you could not drink, did you often have such a strong craving that you could not think of anything else?*
- *Have you had this feeling during the past year?*

0 No undue subjective need.
1 Uneasy, and conscious of need, but not 2.
2 Strong and intrusive preoccupation.

11.008 ***Impaired capacity to abstain or cut drinking (PY and LB)***

Have you recently/ever wanted to cut down or stop using alcohol but could not?

- *Some people try to control their drinking by making rules, like not taking it on weekdays or when alone.*
- *Is that successful for you?*
- *Have you been able to abstain, or to drink at a set lower level for as long as a month?*

0 No difficulty in controlling intake.
1 Has used methods successfully for a month or more.
2 Unable to succeed for as long as a month.

11.009 ***Impaired capacity to control drinking once started (PY and LB)***

Have you found that once you start to drink you take more than you intended?

- *In past year?/Ever?*
- *Have you found it difficult to stop, once started?*
- *Have you ever continued to drink over the whole of a day and into the next, without really being sober? ('Binge' pattern)*

0 No difficulty in controlling once started.
1 Is sometimes successful, sometimes not.
2 Usually takes more than intended once started.
3 Severely impaired control, including 'binges'.

11.010 ***Social problems due to drinking (PY and LB)***

Has drinking led to problems with family, friends or employment? (At school?)

- *Like getting into arguments with spouse or friends; being late at work or missing days; financial problems?*
- *Or more serious problems like losing a friend or partner or being dismissed from work or trouble with the police?*
- *Have any of these happened during the past year?*

0 None.
1 Mild.
2 Moderate.
3 Serious.

11.011 ***Legal problems due to drinking (PY and LB)***

Has drinking led to legal problems on more than one occasion?

- *Like getting into physical fights while drinking or right after drinking?*
- *Getting arrested or held at a police station because of your drinking?*
- *Being taken to court because of physical violence towards your spouse or children?*
- *Being sued due to inability to fulfil financial or role obligations because of your drinking?*
- *Have any of these happened during the last year?*

0 None.
1 Mild
2 Moderate
3 Serious

SKIP => to 11.014 if 11.010 and 11.011 are rated 0.

11.012 ***Persistent drinking after social and/or legal harm (PY and LB)***

Did you continue to drink, even though you knew you had one of these problems?

0 No. 1 Yes.

11.013 ***Failure to fulfil major role obligations (PY and LB)***

When you have been drinking, have you neglected to do things expected of you, like not going to work or school, neglecting to take care of children, or missing important events or appointments?

0 No. 1 Yes.

11.014 ***Risk taking behavior with alcohol (PY and LB)***

[Use culturally appropriate examples.]

Have you ever taken alcohol in situations where it increased your chances of getting hurt?

- *Could you describe that please?*
- *For example, when driving a car, using machinery, or crossing the road?*
- *Have you actually had an accident or been arrested for drunk driving, or hurt yourself?*
- *Have any of these happened during the past year?*

Take into account the degree of tolerance. Assess the risk, not the quantity taken.

0 No risk-taking behavior associated with alcohol.
1 Takes risks involving self and/or others but no harm done.
2 Has been involved in harm to others or to self.

CUT OFF => to 12.001 if heaviest one-day intake ever was less than four standard drinks(men) or three standard drinks (women) AND 11.007 - 11.014 are rated 0 for both periods.

This is a deliberately low threshold. Therefore adapt below-cut-off questions to the period/s of enquiry - past year and/or lifetime before. Use all information from above-cut-off answers to avoid repetitive questioning. Be particularly careful to adapt the questioning if the cut-off intake level has only been marginally passed and no or only minor problems have been indicated.

11.015 ***Salience of drink-related activities (PY and LB)***

How important is drinking compared with other activities?

- *Does it sometimes mean that drinking takes priority over important social obligations?*
- *In past year?/Ever?*
- *For example, do/did you spend money on drink that you should spend on your family?*
- *Or miss appointments because some activity connected with drinking was more important?*
- *Have you given up or reduced participation in important social, recreational or work related activities in order to drink?*

0 No undue priority given to drink-related activities.
1 Some obligations over-ridden but not 2.
2 Drink-related activities nearly always given priority.

11.016 ***Time involved in drink-related activities (PY and LB)***

Have you spent a great deal of time drinking or recovering from the effects of it?

- *In past year?/Ever?*
- *More time than you could afford?*
- *Has there been a period as long as a month when other activities were limited by this?*

Rate the time involved in obtaining, using or recovering from alcohol.

0 No, or only minimal time lost.
1 Much time lost but not 2.
2 Has lost a month or more in more or less everyday involvements.

11.017 ***Narrowing of drinking pattern (PY and LB)***

Have you settled into a routine of drinking?

- *Could you describe the pattern?*
- *Has that changed in any way?*
- *Is it narrower than before; you have to go through the same pattern every day?*
- *Do/did you find it almost impossible to change your habitual routine?*

0 No narrowing of drinking pattern.
1 Some narrowing.
2 Severe narrowing. Pattern fixed in stereotyped routines.

Use all the information available so far about the effects of alcohol on the respondent to phrase the following questions. Tolerance may lessen the physical, psychological or behavioral effects of alcohol.

11.018 ***Tolerance to alcohol (PY and LB)***

Does/did a drink have less of an effect on you than before; you need more to get the same effect?

- *Are you capable of drinking much larger amounts of alcohol than previously without getting the same effects?*
- *How much more?*

0 No development of tolerance.
1 Some but not marked.
2 Marked; can tolerate at least 50% more than previously.
3 Reversed tolerance (see Glossary definition).

11.019 ***Psychological and mental health problems due to drinking (PY and LB)***

Do you think that [Specify most severe type of symptoms.]
is related in any way to drinking?

- *What makes you think that?*

The presence and severity of psychiatric symptoms is rated in detail throughout SCAN, and the relevant sections should be considered before rating this item. A clinical judgement should be based on all the information available, not just the answers to the probes. In the case of psychotic symptoms e.g. alcoholic hallucinosis, an attribution of cause should only be made if there is no or only minor clouding, and onset is typically within 48 hours but up to two weeks of drinking. Do not include symptoms that occur on withdrawal only; they should persist for at least 48 hours but no more than 6 months. Exclude cognitive impairment of various kinds due to alcohol, which are rated at items **21.112, 21.114, 21.116, 21.119** and **21.120**. Mental health problems due to alcohol may also be rated in Section 13 and 20.

00 No symptoms definitely attributable to alcohol.
01 Sleep (S.8) and sexual dysfunction symptoms (S.8)
02 Anxiety (S.4) and obsessive dysfunction symptoms (S.5)
03 Perceptual distortions, etc. (S.16)
04 Depressive symptoms including Psychotic Depression(S.6)
05 Manic symptoms including Bipolar Psychosis(S.10)
06 Delusional symptoms including Morbid Jealousy, but excluding Affective Psychoses and other organic cause(S.19)
07 Hallucinations, particularly auditory, but not organically induced(S.17)
08 Section 18 experiences and/or bizarre delusions.
09 Cognitive impairment (Section 21).
10 Polymorphic symptoms.
11 Mixed symptoms.

SKIP => to 11.023 if 11.019 is rated 0.

11.020 ***Onset of alcohol-induced mental health problems (PY and LB)***

Indicate the circumstances under which symptoms directly attributed to alcohol (item **11.019**) had their onset.

0 No symptoms definitely attributable to alcohol.
1 Onset while drinking or within days of cessation, but not 5 (below).
2 Onset during intoxication, but not 5 (below).
3 Onset while signs and symptoms of physiologic withdrawal in evidence, but not 5 (below).
4 Onset during period characterised by intoxication as well as physiologic signs and symptoms of withdrawal, but not 5 (below).
5 Symptoms persist after all physiologic signs and symptoms of intoxication and withdrawal have abated.
8 Uncertain.
9 NA.

11.021 ***Interference due to alcohol-induced symptoms (PY and LB)***

Consider interference due to symptoms rated at item **11.019**. Exclude social problems and physical health problems rated at items **11.010** and **11.023** as well as neglect of obligations and time lost due to drinking rated in **11.015** and **11.016**. Exclude physiological effects of intoxication and withdrawal.

0 No alcohol-induced mental symptoms.
1 Alcohol-induced symptoms present,but little interference.
2 Moderate or intermittent interference.
3 Severe to incapacitating interference.
8 Uncertain.
9 NA.

11.022 ***Persistent drinking after mental harm (PY and LB)***

Did you continue to drink even when you knew that [problem rated in **11.019**] *was related to alcohol?*

- *When did you know; during the past year or before?*

0 No.
1 Yes.

11.023 ***Physical health problems due to drinking (PY and LB)***

Have you had any health problems due to drinking?

- *What problems? [In past year?/Ever?]*
- *For example, alcohol can cause liver disease, vomiting blood, pancreatitis, pins and needles or numbness in hands and feet.*

Use the CHS to list the ICD-10 categories of any physical illness due to alcohol. If memory impairment, explore connection with drinking but rate in Section 21.

0 No health problems due to drinking.
1 Mild physical illness.
2 Moderate severity.
3 Life threatening.

SKIP => to 11.025 if 11.023 is rated 0.

11.024 ***Persistent drinking after physical harm (PY and LB)***

Did you continue to drink, even when you knew you had one of these health problems?

0 No. 1 Yes.

11.025 ***Alcohol withdrawal problems (PY and LB)***

Have you found, if you cut down or went without alcohol for a time, that you felt any ill-effects?

- *Please describe them.*
- *How often have you had them?*
- *Have any of these been due to cutting down or stopping?*

Check for tachycardia or hypertension, sweating, insomnia, tactile or visual perceptual abnormalities, nausea, diarrhoea, headache, tremor, weakness, restlessness, autonomic anxiety. Recall somatic and neurotic symptoms rated in earlier Sections. Did any of these start hours after withdrawal or cutting down, following a period of heavy drinking? If present at examination, include in past year. Delirium with or without fits is rated 3, and attributed to alcohol at **21.119 or 21.120**.

0 No withdrawal effects.
1 Mild withdrawal effects, monthly or more often.
2 Moderate to severe withdrawal effects (at least 3 symptoms).
3 Delirium or convulsions as withdrawal symptoms.
8,9 As scale I.

11.026 ***Drinking to relieve or to avoid withdrawal symptoms (PY and LB)***

Have you taken alcohol when you had symptoms like that, in order to feel better? Or to avoid these ill-effects entirely?

- *In past year?/Ever?*
- *First thing in the morning?*
- *To stop yourself getting the shakes?*

0 No.
1 Yes, up to once a week.
2 Yes, more often than once a week.

11.027 ***Pathological / idiosyncratic reaction to alcohol (PY and LB)***

Do you find that alcohol has more of an effect on you than on most people?

- *In what way?*
- *How quickly does it start?*
- *Does your personality seem to change when you have had a drink?*

See Glossary definition.

0 No pathological reaction.
1 Definite idiosyncratic or pathological reaction after small doses of alcohol (See Glossary).
2 Inappropriate social behavior only when intoxicated.
3 Violent or aggressive behavior only when intoxicated.
4 Both inappropriate social behavior and violent or aggressive behavior only when intoxicated.
8 Uncertain (Use this rating if description not definite).

11.028 ***Acute toxic effects of alcohol, uncomplicated (PY and LB)***

Were you ever drunk?

- *What was your behavior?*
- *What happened?*

Check for dysfunctional behavior: aggressiveness, impaired judgement and attention, lability; and slurred speech, unsteady gait, conjunctival injection, difficulty standing, flushed face, decreased level of consciousness, nystagmus, without the complications in **11.029 - 11.031,** <u>only</u> when intoxicated..

If present on examination, include in past year.

0 No toxic effects.
1 Mild toxic effects (either type).
2 Moderately severe neurological toxic effect (Glossary).
3 Severe toxic effects, both kinds (but rate delirium, fits, coma at **11.031**).
8 Uncertain.
9 NK.

ACUTE TOXIC EFFECTS OF ALCOHOL, COMPLICATIONS

Medical complications such as trauma, hematemesis, inhalation of vomit, etc. If present at examination include in past year.

SCALE I

11.029 ***Trauma, bodily injury due to alcohol (PY and LB)***

Were you ever injured while drinking?

- *What happened?*

11.030 ***Hematemesis. Aspiration. Other complications (PY and LB)***

Did you ever have any medical problems while drunk?

- *Vomit?*
- *Blood?*
- *Anything else?*

11.031 ***Change of consciousness or perceptual abnormalities due to alcohol (PY and LB)***

Did you ever loose consciousness (use local equivalent term) while drinking?

Did you ever have any strange experiences while drinking? (Use probes from Section 16 and 17 if required).

- *What happened?*

Rate at Section 21, 16, or 17 also.

0 Not present.
1 Perceptual abnormalities present.
2 Delirium.
3 Coma.
4 Associated with convulsions.

11.032 ***Current Drinking***

Are you receiving any help to overcome an alcohol problem?

Use any additional information from informants or records. See item **11.009** for definition of 'binge'.

1 Currently abstinent because institutionalised.
2 Currently abstinent on disulfiram, naltrexone.
3 Currently abstinent under supervisory regime.
4 Currently abstinent not under supervision.
5 Currently using alcohol regularly.
6 Currently using irregularly.
7 Currently drinking in 'binge pattern'.

If necessary rerate **11.006**.

ONSET AND RECENCY OF EPISODES OF DEPENDENCE ON, OR HARM FROM, ALCOHOL

Review with R any positive ratings suggesting dependence or harm, whether during the past year or at some earlier time. Mention the actual items at the points indicated. All items **11.033 - 11.036** should be rated. Rate 9 if NA.

Harm/Abuse items:

Social, legal, mental or physical harm (**11.010, 11.011, 11.019, 11.023**), failure to maintain major role obligations (**11.013**), (**11.014, 11.015**), repeated risk-taking behavior (**11.014**), or continuing to use despite harm (**11.012, 11.024, 11.022**).

The symptoms must continue over a month or more, or occur repeatedly for at least over a year; or for DSM-IV: any time during the same 12 months.

Dependence items:

11.007	-	subjective need.
11.008/11.009	-	inability to control use.
11.015	-	salience of drink-related.
11.016	-	time involved in drink-related activities.
11.018	-	tolerance.
11.012/022/024	-	persistence in use despite evident harm.
11.025	-	withdrawal symptoms.
11.026	-	drinking to relieve or avoid withdrawal symptoms.

Criteria for syndrome of dependence: 3 or more items occurring together at about the same time; either:

- continuously for about a month; or
- repeatedly over about 1 year, or
- DSM-IV: any time in the same 12 month period

11.033 ***Age at first onset of alcohol dependence***

You have mentioned some problems you had with alcohol. For example:
[Specify positive dependence items.]

- *Was there ever a time when several of these problems occurred together at about the same time?*
- *Continuously for about a month or more?*
- *Or on and off over a longer period?*
- *How old were you when that first happened?*

Use the dependence items rated positively during either PERIOD to determine age when a group of such symptoms was first present.

00 = Never present according to criterion.

11.034 ***Age at end of alcohol dependence***

How long did the problems last?
How old were you the last time you had any of them?

Enter age at which the last dependence symptom disappeared. Enter present age if any problem persists at the time of interview.

00 = Never present according to criterion.

11.035 ***Age at first onset of period of harm from alcohol***

[See definition specified above.]

You also mentioned some other problems.
For example: [Specify positive harm items.]

- *When did you first experience this [one of these]?*
- *How old were you then?*

00 = Never present according to criterion.

11.036 ***Age at end of period of harm from alcohol***

How long did [Specify each positive harm item] *last. Is the problem still there?*

Enter age at which last harm problem disappeared. Enter present age if any problem persists at the time of interview.

00 = Never present according to criterion.

Rate interference with activities because of drinking at **13.012**.

11.037 ***Interference with activities because of drinking***

0 No interference
1 Mild interference
2 Moderate interference
3 Severe interference

12 Use of psychoactive substances other than alcohol

The purpose of Section 12 is to determine whether the respondent has used psychoactive substances as prescribed, beyond prescription levels, or completely outside of prescription.

Most items in Section 12 can be rated for two periods, the past year (PY) in the left hand box or upper row of boxes, for each type of drug, and 'lifetime before' (LB) in the right hand box or lower row. The LB ratings are necessary when there have been no problems with drug taking during the past year but there have been problems previously. Specific item-probes for the toxic and withdrawal effects of each class of drug are not included in the text of Section 12 because of the difficulty of presentation. Differential definitions will be found in the Glossary. Specify all drugs used by name.

Use local terms to explain which types of drug are being considered.

Now I should like to ask you about your use of various kinds of drugs. Have you ever used drugs, either on your own or on prescription from a doctor? [Explain drug types if necessary.]

If no:

Not even experimentally, to see what one of them was like?

- *That was the only occasion?*

12.001 ***Lifetime abstention from drugs*** ☐

1 Has never used drugs.
2 Once or twice in a lifetime.

CUT OFF => to 13.001 if 12.001 is rated 1 or 2.

3 Drug/s used more than once or twice.

Continue:

If used on prescription, continue with 12.002.

SKIP => to 12.006 if used other than prescription.

If used on prescription, other than for drug problem, for example, methadone or heroin maintenance, ask:

What sort of drug? Sleeping pills or tranquillizers, or stimulants or diet pills, or painkillers or forms of morphine? [Prompt for type]

- *What problem was it prescribed for?*
- *How often have you taken it during the past year?*
- *Was there an earlier period when you took this or another drug on prescription? During that time, how often did you take it?*

Frequency of use of prescription drug(s):

0 No use.
1 Up to 5 times.
2 6 - 12 times in total, or less than once a month.
3 Approx. monthly.
4 Approx. weekly.
5 Approx. daily.
8 Unsure.

12.002 ***Prescribed Opioids*** ☐

Rate Prescribed Opioids: Morphine, Demerol, Codeine, Methadone, Opium, Darvon, Heroin, Percodan, Dilaudid.

12.003 ***Prescribed Sedatives, hypnotics, anxiolytics*** ☐

Barbiturates, Seconal, Sleeping pills, Tranquillizers, Valium, Librium.

12.004 ***Prescribed Stimulants*** ☐

Amphetamines, Khat, Betel nut, Ritalin.

12.005 ***Use of drugs on and outside prescription***

Did you ever use it in larger amounts or for a longer period, than was prescribed?

- *From friends, street vendors, private suppliers?*
- *Did you ever use other drugs as well as these?*

Include illegal use, culturally sanctioned use, or any means of acquisition other than on legal prescription.

1 Drug/s used only on and according to prescription.
2 Has used prescription drug/s in larger amounts or for longer periods than prescribed.
3 Has used prescription drugs supplied by health professionals and also drugs from illegal sources, including illegal drugs.
4 Illegal use of drug only.

SKIP => to 12.021 if 12.005 is rated 1.

If drugs have been used, or drugs have been taken outside of prescription:

First obtain a general account of the pattern of drug use and any effects or problems, both during the past year and before. Formulate the periods of heaviest and/or most problematic use. The following probes should be used to help respondent give a spontaneous account. Usually only one or a few drugs will have been used. These are designated - [DRUG] or [DRUG/S] - in the text.

What kinds of drug have you used?

- *How do you take drug/s?*
- *By mouth, smoking, sniffing, into a vein or under the skin?*

When did you start using [DRUG/S]? How old were you?
What effects do/did [DRUG/S] have?
Would it be difficult to do without [DRUG/S]?
Has there been any change in pattern of use since then?
Have you ever felt you had problems with [DRUG/S]?

- *Could you describe the problems?*
- *Were the problems worse in the past year or earlier?*

Use this information in subsequent examination. Avoid repetition if later probes have already been answered. The following items should be rated for the past year and/or a previous period of drug use.

Frequency of use of each drug during period:

How often have you used [DRUG/S] during [PERIOD]?

[Prompt from list, **12.006 - 12.014**. If helpful show a prompt card.]

- *As many as 5 times, or up to 12 times, or more than about once a month, or about every week, or more or less daily?*
- *Do you take any of them by injection? (Which most often?)*

0 No use in PERIOD.
1 Up to 5 times in PERIOD.
2 6-12 times in total, or less than once a month.
3 Approximately monthly.
4 Approximately weekly.
5 Approximately daily.
8 Unsure.

12.006 ***Frequency of use of opioids (1) without prescription*** **(PY and LB)** ☐☐

Rate frequency of use of Opioids (1). Morphine, Demerol, Codeine, Methadone, Darvon, Opium, Heroin, etc.

12.007 ***Cannabinoids (2) (PY and LB)*** ☐☐

Ganja, Kif, Marijuana, Hashish, Cannabis.

12.008 ***Sedatives, hypnotics, anxiolytics (3) (PY and LB)*** ☐☐

Barbiturates, Other sedatives, Tranquillizers, Valium, Librium, etc.

12.009 ***Cocaine (4) (PY and LB)*** ☐☐

Coca leaves, Crack.

12.010 ***Stimulants (5) (PY and LB)*** ☐☐

Amphetamines, Khat, Betel nut, Ritalin.

12.011 ***Hallucinogens (6) (PY and LB)*** ☐☐

LSD, Mescaline, Peyote, Psilocybin, DMT, etc.

12.012 ***PCP (7) (PY and LB)*** ☐☐

12.013 ***Volatile Substances (8) (PY and LB)***

Inhalants and Volatile Substances, Glue, Toluene, Petrol/Gasoline Nitrous oxide, Amyl nitrite, etc.

12.014 ***Other and unknown (9)*** **(Specify) (PY and LB)**

The following questions can be rated for 9 types of drug and two periods. The past year (PY) is rated in the upper row of boxes provided for each item, and 'lifetime before' (LB), in the lower row. 'PERIOD' means PY and/or LB, depending on the pattern of use of each drug type. Enter ratings only for relevant drug groups. If not applicable, leave the box blank.

12.015 ***Most recent use (1-9)***

When was the most recent time you used drugs?

- *Even a little bit, or just one time?*
- *How much of the time have you been without it/them since?*

0	No drug use.	5	1 - 6 months.
1	Within past 24 hours.	6	6 months to one year ago.
2	24 - 72 hours.	7	1 + years.
3	4-7 days.	8	Uncertain.
4	Within 1 month.	9	NK, NA.

1.OPD	2.CAN	3.SED	4.COC	5.STM	6.HAL	7.PCP	8.VOL	9.OTH	
☐	☐	☐	☐	☐	☐	☐	☐	☐	PY

12.016.1-9 ***Method of administration during past year (PY) and Lifetime Before (LB)***

0 Method not used.
1 Used, but not the usual method.
2 The usual method of administration.

1.OPD	2.CAN	3.SED	4.COC	5.STM	6.HAL	7.PCP	8.VOL	9.OTH	
☐	☐	☐	☐	☐	☐	☐	☐	☐	PY
☐	☐	☐	☐	☐	☐	☐	☐	☐	LB

12.017.1-9 ***Have you ever used any of these [DRUGS] by injection into a vein?***

1.OPD	2.CAN	3.SED	4.COC	5.STM	6.HAL	7.PCP	8.VOL	9.OTH	
☐	☐	☐	☐	☐	☐	☐	☐	☐	PY
☐	☐	☐	☐	☐	☐	☐	☐	☐	LB

12.018.1-9 ***By injection under the skin or into a muscle?***

1.OPD	2.CAN	3.SED	4.COC	5.STM	6.HAL	7.PCP	8.VOL	9.OTH	
☐	☐	☐	☐	☐	☐	☐	☐	☐	PY
☐	☐	☐	☐	☐	☐	☐	☐	☐	LB

12.019.1-9 ***By smoking?***

1.OPD	2.CAN	3.SED	4.COC	5.STM	6.HAL	7.PCP	8.VOL	9.OTH	
☐	☐	☐	☐	☐	☐	☐	☐	☐	PY
☐	☐	☐	☐	☐	☐	☐	☐	☐	LB

12.020.1-9 ***By snorting or sniffing?***

1.OPD	2.CAN	3.SED	4.COC	5.STM	6.HAL	7.PCP	8.VOL	9.OTH	
☐	☐	☐	☐	☐	☐	☐	☐	☐	PY
☐	☐	☐	☐	☐	☐	☐	☐	☐	LB

12.021.1-9 ***By mouth?***

1.OPD	2.CAN	3.SED	4.COC	5.STM	6.HAL	7.PCP	8.VOL	9.OTH	
☐	☐	☐	☐	☐	☐	☐	☐	☐	PY
☐	☐	☐	☐	☐	☐	☐	☐	☐	LB

Be particularly careful about asking questions **12.022 - 12.027**, if use of drugs has been low and non-problematic. There is no need to ask again if the questions have already been answered in the general enquiry.

Now I should like to ask in more detail about some problems that might have occurred with [DRUG/S] use during PERIOD.

12.022.1-9 *Subjective need for drug*

Have you ever felt that you needed [DRUG] and could not manage without it?

- *How strong is/was that feeling?*
- *Were you uneasy at times or in places when [DRUG] was not available?*
- *In situations where you could not use it, did you ever have such a strong craving that you could not think of anything else?*
- *Have you had this feeling during the past year?*

0 No undue subjective need.
1 Uneasy, and conscious of need, but not 2.
2 Strong and intrusive preoccupation.

1.OPD	2.CAN	3.SED	4.COC	5.STM	6.HAL	7.PCP	8.VOL	9.OTH	
☐	☐	☐	☐	☐	☐	☐	☐	☐	PY
☐	☐	☐	☐	☐	☐	☐	☐	☐	LB

12.023.1-9 *Impaired capacity to abstain or cut drug use*

Have you recently/ever wanted to cut down or stop using DRUG but could not?

- *Some people try to control [DRUG] use by making rules, like not taking it on weekdays or when alone? Is that successful for you?*
- *Have you been able to abstain, or to use it at a set lower level for as long as a month?*

0 No difficulty in controlling intake.
1 Has used methods successfully for a month or more.
2 Unable to succeed for as long as a month.

1.OPD	2.CAN	3.SED	4.COC	5.STM	6.HAL	7.PCP	8.VOL	9.OTH	
☐	☐	☐	☐	☐	☐	☐	☐	☐	PY
☐	☐	☐	☐	☐	☐	☐	☐	☐	LB

12.024.1-9 *Impaired capacity to control drug use once started*

Have you found that, once you have started to use [DRUG] you tend to take more than you intended?

- *In past year?/Ever?*
- *Have you found it difficult to stop, once started?*
- *Have you ever continued to use it over the whole of a day and into the next, without recovering from the effects?*

0 No difficulty in controlling once started.
1 Is sometimes successful, sometimes not.
2 Usually takes more than intended once started.
3 Severely impaired control.

1.OPD	2.CAN	3.SED	4.COC	5.STM	6.HAL	7.PCP	8.VOL	9.OTH	
☐	☐	☐	☐	☐	☐	☐	☐	☐	PY
☐	☐	☐	☐	☐	☐	☐	☐	☐	LB

12.025.1-9 *Social problems due to drug taking*

Has drug taking led to problems with family, friends or employment? (At school or college?)

- *Like getting into arguments with spouse or friends; being late at work or missing days; financial problems?*
- *Or more serious problems like losing a friend or partner or being dismissed from work or trouble with the police?*
- *Have any of these happened during the past year?*

0 None.
1 Mild.
2 Moderate.
3 Serious.

1.OPD	2.CAN	3.SED	4.COC	5.STM	6.HAL	7.PCP	8.VOL	9.OTH	
☐	☐	☐	☐	☐	☐	☐	☐	☐	PY
☐	☐	☐	☐	☐	☐	☐	☐	☐	LB

12.026.1-9 ***Legal problems due to drug taking***

Has drug taking led to legal problems on more than one occasion?

- *Like getting into physical fights while taking drugs or right after taking drugs?*
- *Getting arrested or held at a police station because of your drug taking?*
- *Being taken to court because of physical violence toward your spouse or children under the influence of drugs?*

0 None.
1 Yes. Mild.
2 Moderate.
3 Serious.

1.OPD	2.CAN	3.SED	4.COC	5.STM	6.HAL	7.PCP	8.VOL	9.OTH	
☐	☐	☐	☐	☐	☐	☐	☐	☐	PY
☐	☐	☐	☐	☐	☐	☐	☐	☐	LB

SKIP => to 12.028 if no social and/or legal problems.

12.027.1-9 ***Persistent drug taking after social or legal harm***

Did you continue to take [DRUG], even though you knew you had one of these problems?

0 No.
1 Yes.

1.OPD	2.CAN	3.SED	4.COC	5.STM	6.HAL	7.PCP	8.VOL	9.OTH	
☐	☐	☐	☐	☐	☐	☐	☐	☐	PY
☐	☐	☐	☐	☐	☐	☐	☐	☐	LB

12.028.1-9 *Failure to fulfil major role obligations.*

When you have been using drugs, have you neglected to do things expected of you, like not going to work or school, neglecting to take care of children, or missing important events or appointments?

0 No.
1 Some obligations neglected.
2 Most obligations neglected.
3 Nearly all obligations neglected.

	1.OPD	2.CAN	3.SED	4.COC	5.STM	6.HAL	7.PCP	8.VOL	9.OTH
PY	☐	☐	☐	☐	☐	☐	☐	☐	☐
LB	☐	☐	☐	☐	☐	☐	☐	☐	☐

12.029.1-9 *Risk taking behavior with drugs* [Use culturally appropriate examples.]

Have you ever taken [DRUG/S] in situations where it increased your chances of getting hurt?

- *Could you describe that please?*
- *For example, when driving a car, using machinery, or crossing the road.*
- *Have you actually had an accident or been arrested for driving when under the influence of drugs, or hurt yourself?*

Take into account the degree of tolerance. Assess the risk, not the type or quantity of drug taken.

0 No risk-taking behavior associated with drugs.
1 Takes risks involving self or others but no harm done.
2 Has been involved in risk or harm to others or to self.

	1.OPD	2.CAN	3.SED	4.COC	5.STM	6.HAL	7.PCP	8.VOL	9.OTH
PY	☐	☐	☐	☐	☐	☐	☐	☐	☐
LB	☐	☐	☐	☐	☐	☐	☐	☐	☐

CUT OFF => to 13.001 if most frequent use ever was less than once a month and 12.022 - 12.029 are rated 0 for both periods.

Adapt below cut-off questions to the period/s of enquiry - past year and/lifetime before. Use all information from above cut-off answers to avoid repetitive questioning.

12.030.1-9 ***Salience of drug-related activities***

How important is drug taking compared with other activities?

- *Does it sometimes mean that [DRUG] takes priority over important social obligations?*
- *In past year?/Ever?*
- *For example, do/did you spend money on [DRUG] that you should spend on your family?*
- *Or miss appointments because some activity connected with drugs was more important?*
- *Or have you given up or greatly reduced participation in important social, recreational or work-related activities in order to use drugs?*

0 No undue priority given to drug-related activities.
1 Some obligations over-ridden but not 2.
2 Drug-related activities nearly always given priority.

1.OPD	2.CAN	3.SED	4.COC	5.STM	6.HAL	7.PCP	8.VOL	9.OTH	
☐	☐	☐	☐	☐	☐	☐	☐	☐	PY
☐	☐	☐	☐	☐	☐	☐	☐	☐	LB

12.031.1-9 ***Time involved in drug-related activities***

Have you spent a great deal of time taking or recovering from the effects of [DRUG]?

- *In past year?/Ever?*
- *More time than you could afford?*
- *Has there been a period as long as a month when other activities were limited by this?*

Rate the time involved in obtaining, using or recovering from drugs.

0 No, or only minimal time lost.
1 Much time lost but not 2.
2 Has lost a month or more involved nearly every day.

1.OPD	2.CAN	3.SED	4.COC	5.STM	6.HAL	7.PCP	8.VOL	9.OTH	
☐	☐	☐	☐	☐	☐	☐	☐	☐	PY
☐	☐	☐	☐	☐	☐	☐	☐	☐	LB

12.032.1-9 ***Narrowing of drug taking pattern***

Have you settled into a routine of drug taking?

- *Could you describe the pattern?*
- *Has that changed in any way?*
- *Is it narrower than before; you have to go through the same pattern every day?*
- *Do/did you find it almost impossible to change your habitual routine?*

0 No narrowing of drug taking pattern.
1 Some narrowing.
2 Severe narrowing. Pattern fixed and stereotyped routines.

1.OPD	2.CAN	3.SED	4.COC	5.STM	6.HAL	7.PCP	8.VOL	9.OTH	
☐	☐	☐	☐	☐	☐	☐	☐	☐	PY
☐	☐	☐	☐	☐	☐	☐	☐	☐	LB

12.033.1-9 ***Tolerance to drugs***

Use all the information available so far about the effects of drug taking on the respondent to phrase the following questions.

Does/did a drug have less of an effect on you than before; you need more to get the same effect?

- *Do/did you find you can/could take much larger amounts of [DRUG] than previously without getting the same effects?*
- *How much more?*

0 No development of tolerance.
1 Some but not marked.
2 Marked; can tolerate at least 50% more than previously.
3 Reversed tolerance (See Glossary).

1.OPD	2.CAN	3.SED	4.COC	5.STM	6.HAL	7.PCP	8.VOL	9.OTH	
☐	☐	☐	☐	☐	☐	☐	☐	☐	PY
☐	☐	☐	☐	☐	☐	☐	☐	☐	LB

12.034.1-9 ***Mental health problems due to drug taking***

Do you think that [specify most severe type of symptom] *is related in any way to drug taking? What makes you think that?*

The presence and severity of psychiatric symptoms is rated in detail throughout SCAN, and the relevant sections should be considered before rating this item. The purpose is to specify whether drug taking is the cause of the most severe type of symptom present. (The scale is in order of severity, with 9 the most severe type). A clinical judgement should be based on all the information available, not just the answers to the probes. Psychotic symptoms must have occurred within 2 weeks of drug taking, and persist for at least 48 hours and less than 6 months. Do not include symptoms that occur on withdrawal only. Exclude cognitive impairment due to drugs, which is rated at items **21.113, 21.117, 21.121** and **21.122**. More detailed attributions can also be made using the optional attributional scale and the dashed boxes at an item level, and at the Section level in Sections 13 and 20.

00 No symptoms definitely attributable to drug/s.
01 Impaired judgement or dysfunctional behavior.
02 Anxiety (S.4) and obsessive dysfunction symptoms (S.5) Sleep (S.8) and sexual dysfunction symptoms (S.8)
03 Perceptual distortions, etc. (S.16)
04 Depressive symptoms including Psychotic Depression.
05 Manic symptoms including Bipolar Psychosis.
06 Delusional symptoms including Morbid Jealousy, but excluding Affective Psychoses and rating of (6).
07 Hallucinations, particularly auditory, but not (6).
08 Section 18 experiences and/or bizarre delusions.
09 Cognitive impairment (Section 21).
10 Polymorphic symptoms.
11 Mixed symptoms.

1.OPD	2.CAN	3.SED	4.COC	5.STM	6.HAL	7.PCP	8.VOL	9.OTH	
									PY
									LB

SKIP => to 12.038 if 12.034 is rated 0.

12.035.1-9 ***Onset of drug-induced mental health problems***

Indicate the circumstances under which symptoms directly attributed to drug/s (item **12.034.1-9**) had their onset

0 No symptoms definitely attributed to drug/s.
1 Onset while drugtaking or within days of cessation, but not 5 (below).
2 Onset during intoxication, but not 5 (below).
3 Onset while signs and symptoms of psychological withdrawal in evidence, but not 5 (below).
4 Onset during period characterized by intoxication as well as physiological signs and symptoms, but not 5 (below).
5 Symptoms persist after all physiological signs and symptoms of intoxication and withdrawal have abated. Not taking drugs.
8, 9

1.OPD	2.CAN	3.SED	4.COC	5.STM	6.HAL	7.PCP	8.VOL	9.OTH	
☐	☐	☐	☐	☐	☐	☐	☐	☐	PY
☐	☐	☐	☐	☐	☐	☐	☐	☐	LB

12.036.1-9 ***Interference due to drug-induced symptoms***

Consider interference due to symptoms rated at item **12.034.1-9**. Exclude social problems and physical health problems rated at items **12.025.1-9** and **12.038.1-9** as well as neglect of obligations and time lost due to drugtaking rated in **12.028.1-9** and **12.031.1-9**. Exclude physiological effects of intoxication and withdrawal.

0 No drug-induced mental symptoms.
1 Drug-induced symptoms present, but little interference.
2 Moderate or intermittent interference.
3 Severe to incapacitating interference.
8, 9

1.OPD	2.CAN	3.SED	4.COC	5.STM	6.HAL	7.PCP	8.VOL	9.OTH	
☐	☐	☐	☐	☐	☐	☐	☐	☐	PY
☐	☐	☐	☐	☐	☐	☐	☐	☐	LB

12.037.1-9 ***Persistent drug taking after mental harm***

Did you continue to take drug even when you knew that [problems rated in **12.034**] *were related to [DRUG]?*

- *When did you know; during the past year or before?*

0 No.
1 Yes.

1.OPD	2.CAN	3.SED	4.COC	5.STM	6.HAL	7.PCP	8.VOL	9.OTH	
☐	☐	☐	☐	☐	☐	☐	☐	☐	PY
☐	☐	☐	☐	☐	☐	☐	☐	☐	LB

12.038.1-9 ***Physical health problems due to drug taking***

Have you had any health problems due to drug taking?

- *What problems? [In past year?/Ever?]*
- *For example, DRUG can cause ...*

Specify from Glossary.

Use the Clinical History Schedule to list the ICD-10 categories of any physical illness due to drug taking. Exclude cognitive impairment or decline, which is rated in Section **20**; make the attribution to drugs at **21.113, 21.117, 21.121 or 21.122.**

0 No health problems due to drug taking.
1 Mild physical illness.
2 Moderate severity.
3 Life threatening.

1.OPD	2.CAN	3.SED	4.COC	5.STM	6.HAL	7.PCP	8.VOL	9.OTH	
☐	☐	☐	☐	☐	☐	☐	☐	☐	PY
☐	☐	☐	☐	☐	☐	☐	☐	☐	LB

SKIP => to 12.040 if 12.038 is rated 0.

12.039.1-9 *Persistent drug taking after physical harm*

Did you continue to take [DRUG/S], even when you knew you had one of these health problems?

0 No. 1 Yes.

1.OPD	2.CAN	3.SED	4.COC	5.STM	6.HAL	7.PCP	8.VOL	9.OTH	
☐	☐	☐	☐	☐	☐	☐	☐	☐	PY
☐	☐	☐	☐	☐	☐	☐	☐	☐	LB

12.040.1-9 *Drug withdrawal problems*

Have you found, if you cut down or went without [DRUG/S] for a time, that you felt any ill-effects?

- *Please describe them?*
- *How often have you had them?*
- *Have any of these been due to cutting down or stopping?*

See Glossary for details of the differential withdrawal effects of different drugs. The symptoms should follow withdrawal or cutting down drug intake after repeated or prolonged or high dose use. If present at examination, include in rating of past year. For cocaine and stimulants *dysphoric* mood must also occur. Delirium with or without fits are rated **3**, and attributed to drug at **21.121 - 21.122.**

If necessary, specify from Glossary:

0 No withdrawal effects.
1 Mild withdrawal effects, monthly or more often, or on examination, 1 or 2 effects.
2 Moderately severe withdrawal effects, monthly, or on examination 3 or more effects at a time.
3 Delirium or fits as withdrawal symptoms.

1.OPD	2.CAN	3.SED	4.COC	5.STM	6.HAL	7.PCP	8.VOL	9.OTH	
☐	☐	☐	☐	☐	☐	☐	☐	☐	PY
☐	☐	☐	☐	☐	☐	☐	☐	☐	LB

SKIP => to 12.043 if 12.040 is rated 0.

12.041.1-9 ***Multiple drug withdrawal problems***

Rate here if R has withdrawal symptoms of different drugs listed in the glossary at the same time due to multiple drug use. Use rating scale from previous item.

1.OPD	2.CAN	3.SED	4.COC	5.STM	6.HAL	7.PCP	8.VOL	9.OTH	
☐	☐	☐	☐	☐	☐	☐	☐	☐	PY
☐	☐	☐	☐	☐	☐	☐	☐	☐	LB

12.042.1-9 ***Drug taking to relieve or to avoid withdrawal symptoms***

Have you taken [DRUG/S] when you had symptoms like that, in order to feel better? Or to avoid these effects entirely?

- *In past year/ever?*
- *First thing in the morning?*

0 No.
1 Yes, up to once a week.
2 Yes, more often than once a week.

1.OPD	2.CAN	3.SED	4.COC	5.STM	6.HAL	7.PCP	8.VOL	9.OTH	
☐	☐	☐	☐	☐	☐	☐	☐	☐	PY
☐	☐	☐	☐	☐	☐	☐	☐	☐	LB

12.043.1-9 ***Toxic effects of drugs***

See Glossary for details of dysfunctional behaviors and of the toxic effects of different drugs. If present on examination, include in past year. If delirium or coma, rate 3. Also rate at **21.121 - 21.122** and **21.124**.

0 No toxic effects.
1 Mild toxic effects.
2 Moderately severe neurological effects.
3 Moderately severe dysfunctional behavior.
4 Severe toxic effects, both kinds.
5 Delirium, fits or coma.

1.OPD	2.CAN	3.SED	4.COC	5.STM	6.HAL	7.PCP	8.VOL	9.OTH	
☐	☐	☐	☐	☐	☐	☐	☐	☐	PY
☐	☐	☐	☐	☐	☐	☐	☐	☐	LB

12.044.1-9 ***Current and past drug use***

Are you receiving (have you received) any help to overcome a drug problem?

- *What kind?*

Use any additional information from informants or records.

1 Currently (in past) abstinent because institutionalised.
2 Currently (in past) abstinent on aversive or blocking medication.
3 Currently (in past) abstinent and under supervisory regime.
4 Currently (in past) abstinent not under supervision.
5 Currently (in past) using drug regularly.
6 Currently (in past) using irregularly.

1.OPD	2.CAN	3.SED	4.COC	5.STM	6.HAL	7.PCP	8.VOL	9.OTH	
☐	☐	☐	☐	☐	☐	☐	☐	☐	PY
☐	☐	☐	☐	☐	☐	☐	☐	☐	LB

ONSET AND RECENCY OF EPISODES OF DEPENDENCE ON, OR HARM/ABUSE FROM, DRUGS

Review with R any positive ratings suggesting dependence or harm/abuse, either during the past year or at some earlier time. Mention the actual items, and the relevant drug/s at the points indicated. All items **12.045 - 12.048** should be rated. Rate 9 if NA.

Harm/abuse items:

Social,legal, mental or physical harm (**12.025, 12.034, 12.038**), failure to fulfil major role obligations (**12.028**), repeated risk-taking behavior (**12.029**), or continuing despite harm.

The symptoms must continue over a month, or occur repeatedly over a year, or DSM-IV: any time during the same 12 month period.

Dependence items:

12.022	-	craving.
12.023/12.024	-	inability to control use.
12.027/037/039	-	persistence in use despite evident harm.
12.030	-	salience of drug activities.
12.031	-	time involved in drug-related activities.
12.032	-	Narrowing of drug-taking repertoire.
12.033	-	tolerance.
12.040	-	withdrawal symptoms.
12.042	-	drug-taking to relieve or to avoid withdrawal symptoms.

Criteria for syndrome of dependence; 3 or more items occurring together at about the same time; either:

- continuously for about a month; or
- repeatedly over about 1 year.
- DSM-IV: any time in same 12 month period.

12.045.1-9 ***Age at first onset drug dependence***

You have mentioned some problems you had with DRUG/S, for example: [Specify positive dependence items.]

- *Was there ever a time when several of these problems occurred together at about the same time?*
- *Continuously for about a month? Or on and off over a longer period?*
- *How old were you when that first happened?*

Use the dependence items rated positively during either PERIOD to determine age when a group of such symptoms was first present.

00 = Never present according to criterion.

1.OPD	2.CAN	3.SED	4.COC	5.STM	6.HAL	7.PCP	8.VOL	9.OTH

12.046.1-9 ***Age at end of drug dependence***

How long did the problems last?

- *How old were you the last time you had any of them?*

Enter age at which the last dependence problem disappeared. Enter present age if any problem persists at the time of interview.

00 = Never present according to criterion.

1.OPD	2.CAN	3.SED	4.COC	5.STM	6.HAL	7.PCP	8.VOL	9.OTH

12.047.1-9 ***Age at first onset of period of abuse or of harm from drug taking***

You also mentioned some other problems. For example: [Specify positive harm items.]

- *When did you first experience this [one of these]?*
- *How old were you then?*

See definition specified above.

00 = Never present according to criterion.

1.OPD	2.CAN	3.SED	4.COC	5.STM	6.HAL	7.PCP	8.VOL	9.OTH

12.048.1-9 ***Age at end of period of abuse or of harm from drug taking***

How long did [Specify each positive harm item.] *last? Is the problem still there?*

Enter age at which last harm problem disappeared. Enter present age if any problem persists at the time of interview. Rate interference with activities because of drug-taking at **13.013**.

00 = Never present according to criterion.

1.OPD	2.CAN	3.SED	4.COC	5.STM	6.HAL	7.PCP	8.VOL	9.OTH

12.049.1-9 ***Interference with activities due to drug-taking***

SCALE I

1.OPD	2.CAN	3.SED	4.COC	5.STM	6.HAL	7.PCP	8.VOL	9.OTH	
									PY
									LB

OPTIONAL CHECKLIST (items 12.050 - 12.059).

Use of other drugs not rated in Section 12 (past year only).

0 No use.
1 Moderate ill-effects on health or everyday activities.
2 Serious effects.

12.050 Antidepressants. ☐

12.051 Laxatives. ☐

12.052 Analgesics. ☐

12.053 Antacids. ☐

12.054 Vitamins. ☐

12.055 Diuretics. ☐

12.056 Nicotine dermal patches (number applied per day). ☐

12.057 Steroids / Hormones. ☐

12.058 Specific herbal or folk remedies. ☐

12.059 Other, Specify.. ☐

OPTIONAL APPENDIX CHECKLIST(items 12.060 - 12.105).

Disorders resulting from the use of tobacco (past year only).

Do you smoke, or use tobacco in any other form?

If no:

Have you never smoked or used tobacco?

12.060 ***Lifetime use of tobacco***

1 Has never used tobacco in any form.
2 Once or twice in a lifetime.
3 Tobacco used more than 2.

CUT OFF => to 13.001 if 12.059 rated 1, 2.

Amount of tobacco per average day

How many cigarettes (cigars/pipes/chewing tobacco etc.) have you smoked/used on average, each day during the past year?

Include any form of tobacco in the following ratings, for example: hand rolled cigarettes, bidis, snuff, chewing tobacco, coza/hookah. Estimate amount of tobacco used per day in number if units consumed.

12.061 ***Cigarettes (number of cigarettes)***

12.062 ***Cigars (number of cigars)***

12.063 ***Pipe (number of pipefuls)***

12.064 ***Chewing tobacco (number of bars)***

12.065 ***Other (local terminology, specify)***..

Heavier daily tobacco use

Has there been a period when you used tobacco more heavily than during the past year? How much did you use then?

12.066 ***Cigarettes (number)***

12.067 ***Cigars (number)***

12.068 ***Pipe (number of pipefuls)***

12.069 ***Chewing tobacco (number)***

12.070 ***Other (specify local term)*****..................................**

12.071 ***Record age when started smoking***

How old were you when you started smoking/using tobacco?

- *Do you smoke now?*
- *Do you give up and then start again quite often?*

Make an estimate if Respondent is not sure.

12.072 ***Record age at most recent use***

When did you last give up using tobacco?

SKIP => to 13.001 if Respondent has not used tobacco during past year.

12.073 ***Subjective need for tobacco***

Have you felt that you needed tobacco and could not manage without it?

- *How strong was/is the feeling?*

0 No subjective need for tobacco.
1 Some desire, but can sometimes manage without it.
2 Strong and intrusive preoccupation; craving.

12.074 ***Impaired capacity to abstain or reduce or cut tobacco use***

Have you wanted to stop (smoking) or to cut down, without being able to?

- *Do you make rules to control when you allow yourself to smoke?*
- *Recently, have you been able to go without tobacco for as long as a month?*

0 No difficulty in abstaining or cutting down.
1 Has been successful for a month or more.
2 Unable to succeed as long as a month.

12.075 ***Impaired capacity to control smoking once started***

Have you found that, once started, you tend to continue automatically, without meaning to?

- *Or that, once started, you can't stop?*

0 No difficulty in controlling, once started.
1 Often smokes more than intended, once started.
2 Usually unable to control, once started.

12.076 ***Restriction of social activities due to smoking***

Does smoking lead to social problems with family or friends? (Even to quarrels?)

- *Are any social situations barred because tobacco use is not allowed?*

0 No problems.
1 Moderate problems.
2 Serious problems.

12.077 ***Persistent smoking after social restrictions***

Have you continued to smoke, in spite of these limitations?

0 No.
1 Yes.

SKIP => to 13.001 if respondent uses less than once a week on average, and items 12.072 - 12.076 are rated 0.

12.078 ***Tolerance to tobacco***

Have you increased the amount of tobacco use in order to obtain the same effects?

- *Do you smoke more, or more frequently, to get the same satisfaction?*
- *Do you ever chain-smoke - smoke one after the other?*

0 No development of tolerance.
1 Some tolerance but not marked.
2 Marked tolerance, e.g. chain-smoker.

12.079 ***Restriction of physical activities due to smoking*** ☐

Does smoking prevent you from taking part in physical activity like sports, climbing stairs or anything else?

0 No restrictions.
1 Moderate restriction.
2 Severe restriction.

Health problems due to smoking:

Does smoking have any ill-effects on your health?

- *What are they?*

0 No ill effects.
1 Moderate ill effects.
2 Severe ill effects.

12.080 ***Cough due to smoking.*** ☐

12.081 ***Susceptible to respiratory infections (tobacco).*** ☐

12.082 ***Bronchitis due to tobacco.*** ☐

12.083 ***High blood pressure due to tobacco.*** ☐

12.084 ***Heart trouble due to tobacco.*** ☐

12.085 ***Cancer of the lung due to tobacco.*** ☐

12.086 ***Toxic effects of nicotine (see Glossary).*** ☐

12.087 ***Persistent tobacco use after physical harm*** ☐

Are you concerned about the harmful effects that tobacco might have on your health?

- *Did you continue to smoke when you had health problem and were advised not to?*

0 No.
1 Yes.

Withdrawal problems:

When you reduce or stop smoking [tobacco use] do you get any of the following?

0 No.
1 Yes.

12.088 ***Craving for tobacco.*** ☐

12.089 ***Get irritable and angry (tobacco).*** ☐

12.090 ***Get anxious (tobacco).*** ☐

12.091 ***Have difficulty in concentration (tobacco).*** ☐

12.092 ***Become restless (tobacco).*** ☐

12.093 ***Have increased appetite and gain weight (tobacco).*** ☐

12.094 ***Feel your heart rate slow down (tobacco).*** ☐

12.095 ***Headaches (tobacco).*** ☐

12.096 ***Drowsiness or insomnia (tobacco).*** ☐

12.097 ***Hands shaking (tobacco).*** ☐

12.098 ***Feeling of depression (tobacco).*** ☐

12.099 ***Upset stomach or mouth ulceration (tobacco).*** ☐

12.100 ***Increased cough after cessation.*** ☐

12.101 ***Smoking in order to avoid withdrawal symptoms*** ☐

Do you smoke (etc.) in order to avoid or cure these effects?

0 No.
1 Yes.

ONSET AND RECENCY OF EPISODES OF DEPENDENCE/HARM FROM TOBACCO

Review with R any positive ratings suggesting dependence or harm, whether during the past year or at some earlier time. Mention the actual items at the points indicated.

Harm items:

12.075, 12.078 and **12.079 - 12.085** - social or physical harm; **12.076** and **12.086** - persistence.

The rules require continuous harm for 1 month+, or repeatedly over a year.

Dependence items:

12.072	-	craving.
12.073/12.074	-	inability to control use.
12.077	-	tolerance.
12.076/12.086	-	persistence in use despite evident harm.
12.087-12.99	-	withdrawal symptoms.
12.100	-	smoking to relieve withdrawal symptoms.

Criteria for syndrome of dependence; several of these items occurring together at the same time; either continuously for about a month; or repeatedly over about 1 year.

12.102 ***Age at first onset of tobacco dependence*** ☐☐

You have mentioned some problems you had with tobacco. For example: [Specify positive dependence items.]

- *Was there ever a time when several of these problems occurred together at about the same time? Continuously for about a month or more? Or on and off over a longer period?*
- *How old were you when that first happened?*

Use the dependence items rated positively during either PERIOD to determine age when a group of such symptoms was first present.

12.103 ***Age at end of tobacco dependence***

How long did the problems last? How old were you the last time you had any of them?

Enter age at which the last dependence problem disappeared. Enter present age if any problem persists at the time of interview.

12.104 ***Age at first onset of period of harm from tobacco***

You also mentioned some other problems. For example: [Specify positive harm items.]

- *When did you first experience this [one of these]? How old were you then?*
- *How long did it last?*

See definition specified above.

12.105 ***Age at end of period of harm from tobacco***

How long did [specify each positive harm item] last. Is the problem still there?

Enter age at which last harm problem disappeared. Enter present age if any problem persists at the time of interview.

13 Interference with activities due to Part One symptoms and factors influencing their manifestation (attributions)

CUT OFF => to Section 15 **if the main problems are related to Part Two symptoms only and rate interference, attribution etc. in Section 20.**

CUT OFF => to 13.017 **if there are no or only minimal Section 2 - 12 problems.**

If sections **2-12** have elicited significant problems ask:

13.001 ***Interference with activities due to Part One symptoms***

You have mentioned [summarise symptoms] *during PERIOD. Overall, how much interference has there been with your everyday activities because of these problems?*

Rate interference due to symptoms in Sections **2 - 12**.

0 No Part One symptoms present in significant degree.
1 Symptom present but little interference.
2 Moderate or intermittent interference.
3 Severe to incapacitating interference.

Which symptoms have caused the most interference?

Select the most substantial cause of disablement due to Part One symptoms. Rate only one of the following:

Rate 0 - 3 as **13.001**.

13.002 ***S2.*** ***Physical symptoms*** **(check 2.008).**

13.003 ***S3.*** ***Non-specific symptoms.***

13.004 ***S4.*** ***Anxiety and phobias.***

13.005 ***S5.*** ***Obsessions.***

13.006 ***S6.*** ***Depression*** **(check 6.009).**

13.007 ***S7.*** ***Concentration, energy, etc.***

13.008 ***S8.*** ***Sleep.***

13.009 ***S8.*** ***Sexual problems.***

13.010 ***S9.*** ***Eating disorders* (check 9.005).**

13.011 ***S10.*** ***Mania.***

13.012 ***S11.*** ***Alcohol use.***

13.013 ***S12.*** ***Other substance use.***

13.014 ***S13.*** ***Symptoms associated with trauma or stress.***

13.015 ***Other, specified*** **..**

13.016 ***Unknown or multiple symptoms.***

Hierarchical relationship between Part One symptoms and syndromes:

You have mentioned [summarise symptoms] *during PERIOD. Can you decide which of these types of symptoms or complaints is the most important for you?*

Does any one type of problem (prompt e.g. obsessions, panic, depression, overactivity, drinking, using substances) seem to be the main problem that the others follow from?

If yes:

- *Could you say, if possible, which is next in importance?*

Rate hierarchically which section appears to be primary (rate 1) in Sections 2 - 12. Take account of clinical judgement and all available information. Rate relationship between remaining sections in which items have been rated positively (i.e. code 2, 3 etc. according to order of importance). Give equal ranking to sections that are of equal importance. These ratings, if used, will override prior ratings of syndromal primacy in Sections 2 - 12.

Rate 1, 2, 3 etc:

13.017 ***S2.*** ***Physical symptoms.***

13.018 ***S3.*** ***Non-specific symptoms.***

13.019 ***S4.*** ***Anxiety and phobias.***

13.020 ***S5.*** ***Obsessions.***

13.021 ***S6.*** ***Depression.***

13.022 ***S7.*** ***Concentration, energy, etc.***

13.023 ***S8.*** ***Sleep.***

13.024 ***S8.*** ***Sexual problems.***

13.025 ***S9.*** ***Eating disorders.***

13.026 ***S10.*** ***Mania.***

13.027 ***S11.*** ***Alcohol use.***

13.028 ***S12.*** ***Other substance use.***

13.029 ***S13.*** ***Symptoms associated with trauma or stress.***

13.030 ***Other, specified................................***

13.031 ***Unknown or multiple symptoms.***

13.032 ***Positive activities/enjoyment irrespective of problems***

Most of these questions have been concerned with mental health problems of various kinds. Even if there are such problems it can still be possible for some areas of good functioning to be preserved.

May I ask how far you feel you enjoyed your life during PERIOD and are satisfied with what you achieved during that time?

1 Many positive activities/much enjoyment of life (whether or not there have been symptoms).
2 Some limitation due to Section 2 - 12 problems but areas of activity/enjoyment relatively unaffected.
3 Marked limitation.
4 Serious limitation and relatively little positive experience during PERIOD.

FACTORS INFLUENCING THE MANIFESTATION OF SECTION 2-12 SYMPTOMS DURING THE PERIOD

See Glossary for a more detailed discussion. Both ICD-10 and DSM-III, III-R and IV, require the exclusion of attributions of "cause", particularly in relation to the judged effects of psychoactive substances and alcohol, other treatments, physical (or "organic") disease, and psychosocial stressors and adversity. Users of SCAN who wish to make use of these such classification systems must therefore attempt to consider the grounds for making judgements of this kind. Although opportunity to do so at the item level using the optional attributional scale is present for many items, Section 13 allows the user to re-rate these at a syndromal level, for the purposes of classification, in order to permit the future scientific investigation of the reliability and validity of such attributions.

Note: Users should establish that ratings of attribution can be made reliably. As far as is practicable, such attributional ratings should be delayed until the "causal" status of any attribution can be established with reasonable certainty.

Criteria for rating Attributions of Cause:

(Taken from ICD-10, DCR).

If the following three criteria are met, a *provisional* attribution (i.e. rate 1) can be made:

1 There is objective evidence (from physical and neurological examination and laboratory tests) and/or history of cerebral disease, damage, or dysfunction, or of systematic physical disorder known to cause cerebral dysfunction, including hormonal disturbances (other than alcohol - or other psychoactive substance-related) and non-psychoactive drug effects.

2 There is a presumed relationship between the development (or marked exacerbation) of the underlying disease, damage, or dysfunction, and the mental disorder, the symptoms of which may have immediate onset or may be delayed.

3 There is insufficient evidence for an alternative causation of the mental disorder, e.g. strong family history of a clinically similar or related disorder.

Rate 1, if the above three criteria are satisfied. If not rate 0. If, in addition the following criterion can be met, the attribution can be regarded as *more certain* and rate 2. If in doubt, rate 8. Keep careful records of the evidence upon which ratings are based:

4 There is recovery from or significant improvement in the mental disorder following removal or improvement of the underlying presumed cause.

13.033 ***Quality of data available to rate physical/drug/substance/ organic attribution of cause:***

0 No such attribution, Part One of SCAN.
1 Attribution suspected but inadequate data available.
2 Confident attribution can be based on clinical data.
3 Confident attribution can be based on clinical and laboratory data.

Rate 0, 1, 2, 8, 9 etc:

13.034 ***S2.*** ***Physical symptoms.***

13.035 ***S3.*** ***Non-specific symptoms.***

13.036 ***S4.*** ***Anxiety and phobias.***

13.037 ***S5.*** ***Obsessions.***

13.038 ***S6.*** ***Depression.***

13.039 ***S7.*** ***Concentration, energy, etc.***

13.040 ***S8.*** ***Sleep.***

13.041 ***S8.*** ***Sexual problems.***

13.042 ***S9.*** ***Eating disorders.***

13.043 ***S10.*** ***Mania.***

13.044 ***S11.*** ***Alcohol use.***

13.045 ***S12.*** ***Other substance use.***

13.046 ***S13.*** ***Symptoms associated with trauma or stress.***

13.047 ***Other, specified* ....**

13.048 ***Unknown or multiple symptoms.***

13.049 ***Identify first rated attribution (i.e.) "organic cause" of Part One symptoms.***

Enter ICD-10 Chapter identifier and 1-3 digits. Leave blank if none.
Include dependence-related substances such as alcohol and opioids, etc. These are also rated as causes of abnormal mental states in Sections 11 and 12. Leave blank if none.

13.050 ***Rerate interference if altered by attributed cause:***

13.051 ***Identify second attribution of Part One symptoms.***

If a second attribution of cause is required Enter ICD-10 Chapter identifier and 1-3 digits. Leave blank if none. Include dependence-related substances such as alcohol and opioids, etc. These are also rated as causes of abnormal mental states in Sections 11 and 12. Leave blank if none.

13.052 ***Rerate interference if altered by attributed cause:***

13.053 ***Identify third attribution of Part One symptoms.***

If a third attribution of cause is required Enter ICD-10 Chapter identifier and 1-3 digits. Leave blank if none. Include dependence-related substances such as alcohol and opioids, etc. These are also rated as causes of abnormal mental states in Sections 11 and 12. Leave blank if none.

13.054 ***Rerate interference if altered by attributed cause:***

Individual items can also be rated optionally in more detail at the end of Section **13**, and later at the end of Part Two of SCAN.

The principle underlying the above criteria, appropriately translated, may also be applied to attributions of the effects of therapy, including psychosocial treatment, and the effects of psychosocial stressors. The attribution of abnormal traumas and stressors is required for ICD-10 and DSM (all revisions since III).

For Dissociative Disorders, which may be stress induced, see Section **2**.

These are clinical attributions of cause or precipitation, not simply the respondent's statement.

Items/symptoms should have been rated already in the appropriate sections, apart from a limited number of additional symptoms specific to stress reactions only. In general, these attributions are only required for symptoms that are moderate or severe clinically (i.e. 2 or 3 on Scale I).

ADDITIONAL SPECIFIED FACTORS INFLUENCING THE OVERALL MANIFESTATION OF SYMPTOMS THROUGHOUT PART ONE (SECTIONS 2-4)

Did adverse events or circumstance have any influence or cause the problems you have described (prompt Part One symptoms) during the PERIOD?

- *How important were they?*
- *For example:* [use prompt card if it would be helpful.]
- *Were they the main cause or did they just make things worse or keep them going?*

0 Factor absent.
1 Factor present but effect uncertain.
2 Factor present and modifies the clinical picture or course. Not the main cause or precipitant.
3 Factor the main cause or precipitant and/or markedly modifies clinical picture or course.

13.055 ***Childbirth (only rate 2 or 3 if onset within 28 days), Part One.*** ☐☐

13.056 ***Menopause, Part One.*** ☐☐

13.057 ***Bereavement (only rate 2 or 3 if onset within 3 months), Part One.*** ☐☐

13.058 ***Adolescence, Part One.*** ☐☐

13.059 ***Member of a special cultural group, Part One.*** ☐☐

13.060 ***Recent migration, Part One.*** ☐☐

13.061 ***Life events (use special schedule), Part One.*** ☐☐

13.062 ***Social deprivation, Part One.*** ☐☐

13.063 ***Other, specify, Part One:*** .... ☐☐

For use only after traumatic or stressful experiences. Rate for one period only, which should represent the peak period of illness. If trauma or stressful experience has occurred. Check items **19.061 - 19.075**, which should also be considered following trauma or stress.

The following disorders depend on a detailed and specific clinical history, and on information from case records and informants.

Date onsets at **13.126 - 13.128.**

OPTIONAL CHECKLISTS (ITEMS 13.064 - 13.128)

SKIP => to 13.129 if checklists not used.

STRESS AND ADJUSTMENT DISORDERS

ACUTE REACTION

13.064 ***Sudden overwhelming trauma or stress to R or intimate***: ☐

0 No trauma or events or stressors prior to Part One symptoms.
1 An event outside the range of usual human experience, that would be markedly distressing to almost anyone (i.e. intense fears, helplessness or horror). Events would include catastrophe, serious accident, combat, physical assault, rape, torture or witnessing a violent incident, actual or threatened death.
2 Events within the normal range of human experience, though painful or threatening. Examples include bereavement, natural death, separation, and divorce.

Rate Events at **13.055 to 13.063**.

Specify nature of trauma or stress: ....

SKIP => to 13.129 if 13.064 is rated 0.

13.065 ***Rate period within which symptoms in this section*** **(items 13.067 - 13.098 become positive after stress).** ☐

0 No symptoms.
1 Within 1 hour.
2 Between 1 and 48 hours.
3 More than 48 hours., but less than 4 weeks.
4 Between 4 weeks and 6 months.
5 More than 6 months.

SKIP => to 13.099 If 13.065 is rated 3.

13.066 ***Length of trauma or stress*** ☐

Rate length of trauma or stress. The trauma or stress, of its nature:

1 Is brief - lessens within 8 hours.
2 Begins to diminish within 48 hours.
3 Continues for more than 48 hours.

Items **13.067 - 13.098** are similar to several in Sections **3** and **4** but must have appeared within 48 hours of after **13.064**.

13.067 ***Anxiety or tension related with stress***

Have you felt anxious or restless or tense since [event]?

- *When people get anxious they often feel their heart beating fast, or they start shaking or sweating, or can't get their breath. Or they worry a lot and tend to be irritable and restless. Have you had feelings like that?*
- *Can you describe it?*
- *Or have you felt irritable or unable to sleep or concentrate since?*

0 Anxiety and tension denied.
1 Has had anxiety/tension symptoms.

SKIP => to 13.129 if 13.067 is rated 0.

13.068 ***Duration of symptoms***

How long did these symptoms last?

1 Symptoms began to improve within 8 hours.
2 Symptoms began to improve within 48 hours.
3 No improvement within 48 hours. Symptoms began to improve within one week.
4 Symptoms began to improve within 4 weeks.
5 Symptoms began to improve within 3 months.
6 No improvement within 3 months.
7 No improvement within 6 months.

Rate items **13.069 - 13.098.**

0 absent.
1 Present.

13.069 ***Heart pounding, missing beats, faster.***

13.070 ***Chest pain and discomfort.***

13.071 ***Hot or cold sweats or flushes.***

13.072 ***Trembling of hands or limbs.***

13.073 ***Dry mouth.***

13.074 ***Going crazy.***

13.075 ***Fear of dying.***

13.076 ***Churning stomach, nausea, butterflies.*** ☐ ☐

13.077 ***Can't get breath, choking feeling.*** ☐ ☐

13.078 ***Dizzy, light-headed, faint, unsteady.*** ☐ ☐

13.079 ***Numbness or tingling sensations.*** ☐ ☐

13.080 ***Unreality, 'not really here,' like an actor.*** ☐ ☐

13.081 ***Muscle tension, aches, pains.*** ☐ ☐

13.082 ***Restlessness, inability to relax.*** ☐ ☐

13.083 ***Keyed up, on edge, mental tension.*** ☐ ☐

13.084 ***Difficulty swallowing, lump in throat.*** ☐ ☐

13.085 ***Exaggerated startle reaction.*** ☐ ☐

13.086 ***Poor concentration due to worrying.*** ☐ ☐

13.087 ***Persistent irritability.*** ☐ ☐

13.088 ***Poor sleep because of worrying.*** ☐ ☐

13.089 ***Withdrawal from expected social interaction.*** ☐ ☐

13.090 ***Detachment or estrangement from others.*** ☐ ☐

13.091 ***Narrowing of attention, daze (Inadequate apprehension of external stimuli and events).*** ☐ ☐

13.092 ***Absence of emotional responsiveness (numbness).*** ☐ ☐

13.093 ***Apparent disorientation in time and/or space.*** ☐ ☐

13.094 ***Anger or verbal aggression.*** ☐ ☐

13.095 ***Despair hopelessness, or sense of a foreshortened future.*** ☐ ☐

13.096 ***Inappropriate or purposeless overactivity.*** ☐ ☐

13.097 ***Uncontrollable and excessive grief, judged by local cultural standards.*** ☐ ☐

13.098 ***Dissociative stupor associated with stress.***

Dissociative disorders are rated at **2.053 - 2.070**. See definition of stupor at **2.056**. Rate PSE10 Sections 2 - 6 in PS as soon as possible after stress.

POST-TRAUMATIC STRESS DISORDER

The following items all occur within 6 months of **13.064** and are closely associated with it.

13.099 ***Post-traumatic stress disorder.***

0 Syndrome not present.

SKIP => to 13.115.

1 Onset within 6 months of end of stress.
2 Delayed onset (special projects only).

Rate items **13.100 - 13.114**.

0 Absent.
1 Present.

13.100 ***Significant impairment in social or occupational functioning secondary to the event or symptoms associated with the event.***

13.101 ***Repeated episodes of intensive, painful recollections of the event.***

13.102 ***Avoidance or effort to avoid situations, activities resembling that of stressor.***

13.103 ***Avoidance of associated reminders, thoughts, conversations or feelings.***

13.104 ***Difficulty falling or staying asleep.***

13.105 ***Irritability or bursts of anger.***

13.106 ***Difficulty in concentrating.***

13.107 ***Hyper-vigilance, jumpiness, increased startle response and apprehensiveness, particularly when reminded. Autonomic accompaniments.***

13.108 ***Partial or complete amnesia for the stressful event, which is too extensive and persistent for ordinary forgetfulness.***

13.109 ***Distress at exposure to events (internal or external cues) that symbolize or resemble trauma.***

13.110 Markedly diminished interest in significant activities.

13.111 ***Restricted range of affects (i.e. unable to experience feelings).***

13.112 ***Recurrent distressing dreams, nightmares, with related content.***

13.113 ***Flashbacks: reliving imagery (daytime). Acting on feelings as if events were recurring.***

13.114 ***Acting or feeling as if event were recurring.***

ADJUSTMENT DISORDER

These disorders must follow within one to three months of psychosocial stressor and last less than 6 months. All are covered in detail in PSE10. Rate the predominant emotional type (items **13.119 - 13.125**), then skip to **13.126.**

Stressful life event or significant life change: bereavement, separation, migration, refugee, major change in role: not unusual or catastrophic. Symptoms must be in excess of normal and expectable reaction, and should also be rated in later sections (i.e. Sections 3, 4, 6 etc).

13.115 ***Number of months between stressor and onset (7=7 or more).***

Rate items **13.117, 13.118 - 13.125**.

0 Absent.
1 Present.

13.116 ***Duration less than 6 months.***

13.117 ***Marked distress in excess of response most people would be expected to describe.***

13.118 ***Impairment of social or occupational functioning.***

13.119 ***Brief depressive reaction. Mild. Lasting less than 1 month.***

13.120 ***Prolonged depressive reaction. Mild. 1 month - 2 years.***

13.121 ***Mixed anxiety, and depression.***

13.122 ***Predominant disturbance of other emotions.***

13.123 ***With other prominent specified symptoms.***

13.124 ***Predominant conduct disorder e.g. adolescent grief reaction with aggressive or dystocial behaviour.***

13.125 ***Mixed disturbance of emotions and conduct.***

13.126 ***Age at onset of this episode of Acute Stress Reaction.***

13.127 ***Age at onset of this episode of PTSD.***

13.128 ***Age at onset of this episode of Adjustment disorder.***

ADEQUACY OF INTERVIEW

13.129 ***Adequacy of ratings in sections 2 - 13***

0 No difficulties in making ratings.
1 Some problems but most ratings reasonably accurate.
2 Major difficulty in conducting examination. Ratings must be interpreted with caution.
3 Impossible to rate these sections with any confidence.

Adequacy of Part II is rated at the end of Section **21** and at **24.031**.

Remember to check number and dates of episodes and first onsets in Section 1 on the basis of information provided during the interview.

OPTIONAL RATINGS OF ITEM SPECIFIC ATTRIBUTIONS OF CAUSES AND TRAIT RATINGS

SCAN users and raters with qualification and practice experience in medicine and clinical pharmacology, studying patients following appropriate, extensive, physical and laboratory investigation may wish to rate judgements of definite attributions of cause at an item level. Only attributions that can be positively rated should be entered. For each period first rate the item using the appropriate SCAN rating scale (I-IV), in the standard scoring boxes, and then the nature of the attribution in the dashed boxes below the standard boxes, using the scale provided. Up to two periods can be rated. Note that the conventional use of 0,8, and 9 does not apply and these scores have specific meaning. The items themselves already should have been rated purely on the basis of certainty, presence, and severity.

The following optional ratings should only be made by investigators planning to study their validity. The reliability of making these ratings should be established. They represent a more detailed way of recording ratings of attribution of organic cause throughout the SCAN. There is also an opportunity to make a trait rating of individual items (i.e. a rating of "7" on rating SCALE I of SCAN version 1). Users should contact their training center for up to date information of their employment in SCAN algorithms.

ATTRIBUTIONAL RATING SCALE (Optional)

Use the attributional rating scale to rate nature of influence on symptom presence or severity at the item level. The rating of 0 is the only 'negative' rating, and can be used if there is clear evidence that treatment has resulted in a reduced rating. Ratings 1 - 8 represent attributions of etiologic effect, increasing the likelihood of an item being rated or its severity being increased. These attributions must only be made for symptoms/items that have positively rated, and should be entered into the dashed boxes below the standard episode rating boxes. The rating of 9 is reserved for indicating a trait characteristic. If no effect or trait is judged to be present leave the dashed box blank.

0 Item rating reduced by effects of treatment
1 Alcohol
2 Other psychoactive substance
3 Effects of somatic psychiatric treatments (electroconvulsive therapy, antidepressant medication, neuroleptics, etc)
4 Known primary intracranial process (Alzheimer's disease, Huntington's disease, Parkinson's disease, tumor, stroke, etc.)
5 Non-psychiatric medication, toxins
6 Other medication condition 1 (specify at Section 13 and/or Section 20)
7 Other medication condition 2 (specify at Section 13 and/or Section 20)
8 Other medication condition 3 (specify at Section 13 and/or Section 20)
9 Trait. Essentially lifelong characteristic of R.

14 Screen for items in Part Two

SKIP => to Section 15 if Part Two is used. This screen is optional.

This screen is composed of items relevant to items in Part Two, which is a separate schedule and not to be included with Part One, except when the purposes of the investigator require it. The screen may be useful also when both parts are to be used, but then any positive screen items should be rated on the full scale in the appropriate section. It is essential to record R's description of any positive symptoms.

14.001 ***Change in appearance of things***

Some people occasionally get a feeling that the appearance of things, or people, or even themselves, has changed. That things look or sound or smell unusual or that time has become distorted. Have you had such feelings in [PERIOD]?

Probe as necessary then rate.

0 No.
1 Yes.

14.002 ***Delusional mood and perplexity***

Have you had the feeling that something odd is going on that you can't explain?

- *What is it like?*
- *Do you feel puzzled by strange happenings that are difficult to account for?*

R feels that the familiar environment has changed in a way that puzzles her/him and which she/he may not be able to describe clearly. The feeling often accompanies delusion formation.

0 No.
1 Yes.

14.003 *Interference with thoughts*

Can you think quite clearly, or does there seem to be some kind of interference with your thoughts?

- *What is that like?*
- *Are you fully in control of your thoughts and actions?*

Probe as necessary then rate. Exclude slowness or muddled thinking e.g. in depressive states.

0 No.
1 Yes.

14.004 *Second sight, strange presences*

What about other unusual experiences that some people have, such as seeing things that others cannot see, having second sight, or being aware of strange presences?

- *Can you describe such an experience?*

Probe as necessary then rate:

0 No.
1 Yes.

14.005 *Hearing voices*

We ask this question of everyone and would like to ask you. Do you ever seem to hear noises or voices when there is nobody about and no ordinary explanation seems possible?

- *What is that like?*

0 No.
1 Yes.
8 Unsure.

14.006 ***People too interested in R***

Have you had a feeling that people were too interested in you?

- *Or that things were arranged so as to have a special meaning for you, or even that harm might come to you?*
- *Can you describe that?*

Probe as necessary then rate:

0 No.
1 Yes.

14.007 ***Odd or unpleasant experiences***

Have there been any other odd or unpleasant experiences of any kind recently?

- *What happened?*

Probe as necessary then rate:

0 No.
1 Yes.

14.008 ***Subjectively described memory problems***

Have you had any difficulty with your memory?

- *Is it more difficult to remember things than it used to be?*
- *Can you give me an example?*
- *When did the problem start?*

Do not include trait inattention or 'absent-mindedness'. There must be a definite loss of memory function, though not necessarily with onset in [PERIOD].

0 None.
1 Mild difficulty such as 'forgetfulness' that might be due to inattention, worry, etc.
2 Serious memory loss, unlikely to be due solely to inattention, worry, etc.

Record examples.

14.009 ***Compensation for memory impairment***

Do you have to keep notes to remind you of things to be done?

There must be a definite change, though not necessarily with onset in [PERIOD].

0 None
1 Mild difficulty that might be due to inattention, worry, etc.
2 Serious memory loss, unlikely to be due solely to inattention, worry, etc.

Thank you very much for your help. I hope the questions did not worry you. Most people are quite interested to answer them. Have you any comments or questions?

Rate the following items from Sections 22 - 24 but rate the full sections if this selection d oes not cover all the signs present.

0 Symptoms did not occur during period of observation.
1 Definitely present in moderate degree.
2 Present in severe degree.

14.010 ***Slowness***

Very slow to move. Unusual for age and physical condition. Motor retardation.

14.011 ***Restlessness***

1 Noticeably restless, not amounting to 2.
2 Constantly fiddling, changing position, standing or sitting down, etc.

14.012 ***Odd or inappropriate appearance***

Odd clothes, ornaments etc. Would look odd due to posture, gait etc.

14.013 ***Self neglect***

Clothes inadequate for warmth and protection. Irrespective of whether odd or embarrassing.

14.014 ***Observed anxiety***

Fearful apprehensive look, frightened tone of voice, tremor in voice/hands, autonomic signs.

14.015 ***Observed depression***

Sad, mournful look, tears, gloomy tone of voice, deep sighing, voice chokes on distressing topic.

14.016 ***Blunting or flattening of affect***

Decrease in emotional responsiveness shown in facial expression, tone of voice, use of gesture, etc.

1 Blunting. Quantitative decrease compared with expectation.
2 Severe and uniform flatness of affect.

14.017 ***Incoherence of speech***

The subject's meaning is obscured by distorted grammar, lack of logical connection between one part of a sentence and another or between sentences, sudden irrelevancies or 'Knight's move', incomprehensible associations, etc.

Do not rate this symptom present unless examples are written down.

14.018 ***Magical or markedly illogical thinking***

14.019 ***Rate restricted quantity of speech***

Subject frequently fails to answer, questions have to be repeated, restricted to minimum necessary, no extra sentences, no additional comments.

14.020 ***Poor use of non-verbal communication during interview***

Lack or under-use of normal non-verbal gestures, facial expressions, changes of tone and pitch and loudness, eye- contact, etc., during conversation.

If Part Two and IGC are not used.

CUT OFF => to Clinical History Schedule, Section 27.

PSE Part Two

15 Language problems at examination

When rating items in Sections 16 - 19 remember that any serious language difficulty, from whatever cause, imposes serious limitations on the respondent's understanding of questions and on the interviewer's interpretation of answers. For valid rating, R must be able to understand the questions and clearly describe the relevant experiences without undue prompting.

If cognitive or language problems make responses to questions in Sections 16 - 19 problematic, rate 5 on Scale II.

Rate language problems:

0 None.
1 Items in Sections 16 - 19 can be rated but must be interpreted with caution.
2 Difficulty so marked that ratings are not applicable.

15.001 ***Poor language ability due to limited intelligence*** ☐

15.002 ***Incoherent speech from any cause*** ☐

15.003 ***Developmental language disorder of autistic spectrum*** ☐

Rate here the concreteness, repetitiveness, dominance of a single theme, and otherwise poverty of content of Asperger's syndrome, as well as more severe autism.

15.004 ***Other developmental language disorder, e.g. dysphasia*** ☐

15.005 ***Poverty of content or too little speech*** ☐

15.006 ***Disturbance of consciousness*** ☐

15.007 ***Poor comprehension and/or use of English, if not R's mother tongue*** ☐

15.008 ***Other reasons for language deficit*** ☐

Specify:..

SKIP => to 21.065 if any of 15.001 - 15.008 are rated 2.

If evidence of autistic spectrum, developmental disorders should also be rated in CHS, see items 27.012 - 27.019.

RATING SCALE II

NB: Consider behavioural items for cognitive impairment, **21.065 - 21.088**.

Most items in Sections 16 - 19 cover psychotic items that can be rated in terms of frequency or of duration. Choose whichever is more appropriate.

0 Symptom did not occur during PERIOD.
1 Symptom definitely occurred during the period but were probably uncommon or transitory.
2 Symptom was definitely present, on multiple occasions or for part of the time, during the period.
3 Symptom was present more or less continuously throughout the period.
5 Language difficulty, rated as present in Section 15, makes replies difficult to interpret.
8 Rater is unsure whether the phenomenon is present or absent, even after adequate examination.
9 Not appropriate to make a rating because examination incomplete, e.g. because of refusal, omission, etc.

NB: In the case of delusions, the rating is made on the approximate length of time R was deluded during the PERIOD and on the continuity of the beliefs. A rating of 2 applies only if R is not continuously deluded.

16 Perceptual disorders other than hallucinations

16.001 ***Unusual sensations***

Do things look and sound and feel normal or have you been aware of unusual sensations?

- *Some people occasionally experience a feeling that things or people are not real.*
- *Does time seem to pass normally?*

Probe as necessary.

0 No.
1 Yes.

CUT OFF => to 17.001 if no evidence for Section 16 Symptoms.

16.002 ***Changing perceptions***

For example, do things seem to change in size or shape or color in a puzzling way?

- *What is that like?*

0 - 3 Use Scale II

16.003 ***Dulled perception***

Have things looked grey and flat; lacking their usual color and detail?

- *Can you describe that?*

0 - 3 Use Scale II

16.004 ***Heightened perception***

Do sounds seem unnaturally clear or loud or objects look vividly coloured or patterns seem particularly detailed and interesting?

0 - 3 Use Scale II

16.005 ***Changed perception of time, déja vu, jamais vu, etc.***

Does your experience of time seem to have changed?

- *Does it go too fast or too slowly or do you seem to live through events exactly as you have had them before?*

0 - 3 Use Scale II

16.006 ***Derealisation (things)***

Have you felt recently as though the world was unreal, or only an imitation of reality, like a stage set, with cardboard cut-outs instead of real houses or trees?

- *What was that like?*

0 - 3 Use Scale II

16.007 ***Derealisation (people)***

Did other people seem to be acting a part, like actors in a play, or like puppets, or even dead?

0 - 3 Use Scale II

16.008 ***Depersonalization***

Have you felt that you yourself were not a real person, not really part of the living world?

- *Like being in a dream? "Not really here"?*
- *Like acting in a play with all the lines laid down?*

0 - 3 Use Scale II

16.009 ***Depersonalized perception of self***

Do you seem unreal to yourself when you look in a mirror?
Do you find that you seem to be seeing yourself from outside your body, like a stranger?

0 - 3 Use Scale II

16.010 ***Unfamiliarity (self)***

Have you felt that part of your body did not belong to you or looked unfamiliar, or the wrong size?

0 - 3 Use Scale II

16.011 ***Changed personal appearance (dysmorphophobia)***

Does your appearance seem to have changed recently? Are your features or body the same as usual? (Nose, teeth, mouth, etc.).

- *Is there really a change that other people can see, or is it just a feeling*

Rate non-delusional dysmorphophobia, i.e. R accepts that there is not really any distortion, but still has the feeling; rate preoccupation at **2.038** and conviction at **2.039**.

0 - 3 Use Scale II

16.012 ***Delusions concerning appearance***

Include dysmorphophobic delusions check **19.031**.

0 - 3 Use Scale II

16.013 ***Delusion of depersonalization or derealisation***

Do you think that some part of your body or mind is missing - like no head, no brain, no thought, or no mind?

Include any delusional form of depersonalization, not only the delusion of non-existence, named after Cotard. Check at item **19.032**.

0 - 3 Use Scale II

16.014 ***Perceptual abnormalities nos.***

0 - 3 Use Scale II

Record description.

16.015 ***Dates of PERIOD of Section 16 symptoms***

It is necessary to record only if dates of symptoms in this Section are different from PERIOD rated in Section 1 (items **1.007**, **1.009**, **1.010 - 1.013**).

16.016 ***Interference with activities due to Section 16 symptoms***

You have mentioned [summarise symptoms] during PERIOD. Overall, how much interference has there been with your everyday activities because of these problems?

Rate interference due to symptoms in Section 16.

0 No Section 16 symptoms present in significant degree.
1 Symptom/s present but little interference.
2 Moderate or intermittent interference.
3 Severe to incapacitating interference.

Although opportunity to rate attribution of physical (including drug, substance etc) cause is present using the optional attributional scale and the dashed boxes, items below allow the user to re-rate organic cause at the section / syndrome level.

16.017 ***Organic cause of Section 16 symptoms***

Use the four criteria listed in Section 13.

0 Absent.
1 Probable organic cause; not fully confirmed
2 Definite, confirmed by expert investigation.
8 Uncertain whether organic or not.

16.018 ***Identify the organic cause of Section 16 symptoms***

Exclude conditions rated in Sections 11, 12 and 20 as causes. For other conditions, enter identifying letter for ICD-10 chapter and up to 3 digits, here at **16.018**. Leave blank if none.

17 Hallucinations

17.001 ***Probe for hallucinations***

We ask this question of everyone and would like to ask you. Do you ever seem to hear noises or voices when there is nobody about and no ordinary explanation seems possible, or see or feel things other people cannot?

- *What is that like?*

0 Hallucinations absent.
1 Hallucinations present.
8 Unsure.

17.002 ***Probe for other unusual experiences***

Have there been any other odd or unpleasant experiences of any kind recently?

0 Absent.
1 Present.

CUT OFF => to 18.001 if no evidence of hallucinations.

17.003 ***Non-verbal auditory hallucinations***

Do you hear noises like music or birds or muttering or whispering?

- *Can you describe it?*
- *Can you make out if there are any words?*
- *How often?*

[Exclude tinnitus, etc.]

0 - 3 Use Scale II

SKIP => to 17.015 if no auditory hallucinations.

17.004 ***Frequency of verbal auditory hallucinations (AH)***

How often do you hear it/them?

- *Rarely, every week or so, every day, most of the time?*
- *Has there been a time during [PERIOD] when you were free for at least a week .*

[Include all forms of AH in this rating.]

1	Rarely.	3	Most days.
2	Every week or so.	4	Nearly all the time.

17.005 ***Length of AH utterances***

What does it (they) say?

- *Can you give me some examples?*
- *Do they just say a few words or is there a long monologue (or conversation between voices)?*

If R says voices go on for a long time:

Are they just repeating the same brief sentences over and over?

[Include all forms of AH in this rating.]

0 No AH.
1 A few words or brief sentences only.
2 Mostly few words but occasionally longer continuous utterances. Not just repetitions.
3 Long, more or less continuous AH.
4 Threatening or commanding voices.

17.006 ***Quality of AH***

What are the voices like? Are they like a real voice? Can you tell them from my voice, for example?

- *Is there a special quality to them? What is it like*?

0 No AH.
1 Special quality; not much like real voices.
2 Sound more or less like real voices.

17.007 ***Internal hallucinations***

Do you hear them in your head or mind, or in your ears, or as though coming from outside?

- *Where do they seem to come from?*

Inner voices or images, perceived with the vividness and concreteness characteristic of hallucinations, but lacking external projection. Insight into the morbid nature of the experience may or may not be present; it is not a defining feature.

0 No AH.
1 Predominantly external hallucinations.
2 Partly within head/mind, partly outside.
3 Predominantly inner hallucinations.

17.008 ***Voice(s) commenting on thoughts or actions***

Does a voice comment on your thoughts?
Does a voice repeat things you have thought?
Do you hear a voice saying what you are reading, or describing what you are seeing on television as you see it?

- *How often does it happen?*

0 No AH.
1 No interaction between voices and thoughts.
2 Voices do comment or repeat or describe, occasionally.
3 Frequent comments or repetitions of thoughts or actions.

17.009 ***Second and/or third person auditory hallucinations***

Do you hear voices talking to each other or directly to you?

- *What do they say to each other?*
- *Do they talk about you between themselves?*
- *Do you ever hear a single voice talking about you?*

What about a voice or voices talking directly to you?

If both:

Which kind of voice is commoner - the one talking to you or the one talking about you?

0 No AH.
1 Second person only.
2 Both second and third person, but more second than third.
3 Neither more prominent than the other.
4 Both second and third person, but more third than second.
5 Third person only.

17.010 ***Congruence of AH with affective state***

Rate congruence: for example, content based on guilt: "*You deserve to die you sinner*", etc., or on delusions of grandeur: "*Go to the palace. You are the King*", etc. Include if content was congruent earlier but mood has improved before AH disappeared. If unsure, rate 8. Check at **6.021**.

0 No AH.
1 Virtually all content congruent with affective state.
2 More congruent than not.
3 Congruent and incongruent AH equally.
4 More incongruent than congruent.
5 Virtually all content incongruent.
9 No affective state.

17.011 ***Mood rated congruent at item 17.010***

0 No or very little congruence.
1 Congruence is mainly with depressed mood.
2 With both depressed and elated mood.
3 Mainly with elated mood.

17.012 ***Special features of auditory hallucinations***

Are there any other characteristics of the voices?

- *Do you hear them only through other noises? For example, some people hear voices only through fan or aeroplane noises or in the cries of birds, and not if it is quiet.*
- *Do you hear the voice from a part of your body?*
- *Does the voice ever come out of your own mouth?*

0 None present.
1 Special features present.

17.013 ***Insight into auditory hallucinations***

How do you explain the voices? Where do they come from?

- *Why do you hear them?*

0 No AH.
1 Explanation in natural terms - '*my thoughts*', '*illness*'.
2 No explanation. '*Don't know*'. Puzzled.
3 Explanation without insight, e.g. delusional.

17.014 ***Prominence of auditory hallucinations in total clinical picture***

0 No hallucinations.
1 Hallucinations present but not acentral feature of the clinical picture.
2 Hallucinations are a prominent feature of the clinical picture but not 3.
3 AH vigorously intrusive, and persistent compared with other symptoms.

Dissociative auditory hallucinations rate with similar visual hallucinations at **17.020**.

If organic cause for hallucinations, use etiology option or rate at 17.034-17.035.

VISUAL HALLUCINATIONS

First obtain and record in R's words a description of the phenomena.

Have you had visions or seen things other people couldn't? What did you see?

- *Was it flashes or shadows, or formed people or objects?*
- *Was it whole scenes or only particular people or objects? (With your eyes or in your mind?)*
- *Were you half-asleep at the time?*
- *Has it occurred when you were fully awake?*
- *Did you think the visions were real?*

If a person:

Did you recognise the person? Who was it?

- *Did s/he say anything? What?*
- *Could you hold a two-way conversation?*
- *Do you know anyone who has had this kind of experience?*

Skip => to 17.022 if no visual hallucinations.

17.015 ***Unformed visual hallucinations, clear consciousness***

Exclude organic cause and drug effects.

0 - 3 Use Scale II

17.016 ***Formed visual hallucinations, not whole scenes***

0 - 3 Use Scale II

17.017 ***Scenic visual hallucinations in clear consciousness***

0 - 3 Use Scale II

17.018 ***Hypnogogic and hypnopompic hallucinations***

0 - 3 Use Scale II

Rate visual hallucinations with clouded consciousness at **21.070**.

17.019 ***Brief auditory and/or visual hallucinations associated with bereavement***

The most typical example is seeing and/or hearing a recently dead person in some very familiar setting. It could also be someone who has gone away. A familiar remark, usually brief, may be heard quite clearly, or a familiar face may be seen momentarily, e.g. in a mirror. Exclude cultural or religious experiences.

0 - 3 Use Scale II

17.020 ***Dissociative hallucinations***

The typical example is someone who can see and/or carry on a two-way conversation with another, usually known person. Sometimes there is a special occult context, as when R is a member of a cult or family with such beliefs or practices, or has been in the past. TV programmes or books may be another influence. If the experience occurs as part of a social or cult ceremony or ritual, consider items **2.053 - 2.070** and **19.021 - 19.024**.

0 - 3 Use Scale II

17.021 ***Prominence of VH in total clinical picture***

0 No visual hallucinations.
1 VH present but not a central feature of clinical picture.
2 VH a prominent feature of the clinical picture but not 3.
3 VH dominate the clinical picture compared with other symptoms.

Rate VH with altered consciousness at **21.070**.

OTHER HALLUCINATIONS AND SOMATIC DELUSIONS

Skip => to 17.031 if no other hallucinations.

These hallucinations and the delusional elaborations and explanations to which they give rise are placed together for convenience. Other delusions, not clearly associated with hallucinations are in Section 19.

17.022 ***Olfactory hallucinations***

Have you noticed unusual smells that you cannot account for?

Do not rate delusional explanations, only the hallucination, if present. Exclude **17.023 - 17.025**.

0 - 3 Use Scale II

17.023 ***Bizarre delusions associated with smell***

*What is the explanation (for **17.022**)?*

['Gas pumped into room', etc.]

0 - 3 Use Scale II

Rate only any delusional element. Exclude **17.024 - 17.025**.

17.024 ***Hallucination that R gives off a smell***

Do you think that you yourself give off a smell?

- *Even when you know you are quite clean?*
- *Can you describe what it is like?*
- *What is the explanation?*

May be associated with **17.025** but not necessarily.

0 - 3 Use Scale II

17.025 ***Delusion that others think R smells***

Do other people think you give off a smell?

- *Even when you do not?*
- *How do they show this? What do you notice?*
- *How do you explain it?*

17.024 may be denied, although **17.025** is present

0 - 3 Use Scale II

17.026 ***Sexual hallucinations***

Do you have any unusual sexual sensations?

- *Can you describe them?*
- *How do you explain them?*

Rate only the hallucination, if present. Any delusional explanation is rated at **17.027**.

0 - 3 Use Scale II

17.027 ***Sexual delusions associated with hallucinations***

How do you explain these sensations?

Sexual feelings induced by telepathy, bizarre explanations, changing sex. Other sexual delusions, not apparently based on hallucinations, should be rated at **19.014 - 19.018**.

0 - 3 Use Scale II

17.028 ***Hallucinations of other senses***

Do you notice other strange or inexplicable sensations of touch, or temperature, or pain, or floating? Or like a crawling sensation under the skin?

- *Can you describe it?*

0 - 3 Use Scale II

Rate only the hallucination, not any delusional elaboration or explanation.

17.029 ***Bizarre delusions associated with somatic sensations***

*What is the explanation (for **17.028**)?*

'Torture', 'ice instead of heart', 'insects inserted under skin', 'floating in air or flying', 'organs misplaced', 'liver made of gold', 'animals passing through body', etc.

0 - 3 Use Scale II

17.030 ***Apparent hallucinatory experiences nos.***

Include occult or inexplicable experiences not associated with group beliefs or practices.

0 - 3 Use Scale II

17.031 ***Enter age at first onset of hallucinations***

Check date at **1.013** and **1.017**.

17.032 ***Dates of PERIOD of Section 17 symptoms PE*** to

It is necessary to record only if symptoms in this Section are different from PERIOD rated in Section 1 (**1.007**, **1.009**, **1.010 - 1.013**).

17.033 ***Interference with activities due to Section 17 symptoms***

You have mentioned [summarise symptoms] during PERIOD. Overall, how much interference has there been with your everyday activities because of these problems?

Rate interference due to symptoms in Section 17.

0 No Section 17 symptoms present in significant degree.
1 Symptom/s present but little interference.
2 Moderate or intermittent interference.
3 Severe to incapacitating interference.

Although opportunity to rate attribution of physical (including drug, substance etc) casue is presentusing the optional attributional scale and the dashed boxes, item **17.034**and **17.035** allows the user to re-rate organic cause at the section/syndrome level.

17.034 ***Organic cause of Section 17 symptoms***

0 Absent.
1 Probable organic cause not fully confirmed.
2 Definite, confirmed by expert investigation.
8 Uncertain whether organic or not.

17.035 ***Identify the organic cause of Section 17 symptoms***

Include conditions rated in Sections 11, 12 and 21 as causes; e.g. attribute alcoholic hallucinosis at 11.019 and in Section 20. For other conditions causing hallucinations record only in Section 20. At **17.035**, enter identifying letter for ICD-10 chapter and up to 3 digits. Leave blank if none.

Remember to complete Section 20 history items (for psychotic and affective disorders, if relevant.

18 Subjectively described thought disorder and experience of replacement of will

If speech or language disorder is present, rate Section 18 items present only if it is quite clear that R understands the symptom and describes it without echoing leading questions. A simple *'yes'* in answer is not evidence.

18.001 ***Delusional mood and perplexity***

Have you had the feeling that something odd is going on that you can't explain?

- *What is it like?*
- *Do you feel puzzled by strange happenings that are difficult to account for?*
- *Do familiar surroundings seem strange?*

R feels that the familiar environment has changed in a way that puzzles her/him and which she/he may not be able to describe clearly. The feeling often accompanies delusion formation. Rate only when mood precedes delusion formation and do not rate present if perplexity occurs on its own (or rate 8).

0 - 3 Use Scale II

18.002 ***Probes for section 18 symptoms***

Can you think quite clearly, or does there seem to be some kind of interference with your thoughts?

- *What is that like?*
- *Are you fully in control of your thoughts and actions?*

0 Section 18 symptoms absent (other than **18.003**).
1 Section 18 symptoms present (other than **18.003**).

CUT OFF => to 19.001 if 18.001 and 18.002 are rated 0.

18.003 ***Delusion that thoughts are being read***

Has it seemed that your thoughts were read by other people?

- *Can you describe that?*

0 - 3 Use Scale II

NB: This item is included here for convenience of interviewing. See Glossary for differential definition from other items in Section 18.

18.004 ***Loud thoughts***

Do your thoughts seem to sound aloud in your head, almost as though someone standing near you could hear them?

- *What is that like?*

Please read Glossary definitions carefully for all Section 18 symptoms.

0 - 3 Use Scale II

18.005 ***Thought echo***

Does a thought in your mind seem to be repeated over again, like an echo? Can you describe it for me?

- *What is it like?*

'Voices repeating thoughts' should be rated at **17.008**.

0 - 3 Use Scale II

18.006 ***Thought insertion***

Do there seem to be thoughts in your mind which are not your own; which seem to come from elsewhere?

- *How do you think they get in your mind?*

0 - 3 Use Scale II

18.007 ***Thought broadcast***

Do your thoughts seem to be somehow public; not private to yourself, so that others can know what you are thinking?

The experience of thought broadcast must be described. Exclude delusion that thoughts are subsequently used on TV, in newspapers, etc. See item **19.006**.

0 - 3 Use Scale II

18.008 ***Thought commentary***

Does there seem to be another stream of thoughts in your mind, not under your control, which might, for example, comment on your thoughts, or on something you are reading or something you have seen or done?

- *Is that like a voice or is it another kind of thought?*
- *What is that like?*

Exclude AH comments in this rating.

0 - 3 Use Scale II

18.009 ***Thought block***

Do your thoughts sometimes stop suddenly, so that your mind is a complete blank, although you have not yourself wanted to stop thinking?

- *Can you describe that?*
- *When it stops, do you pick up your thoughts where they left off?*

Differentiate from lapse of attention, or distraction or anxiety. There must be absence of thought, not wandering.

0 - 3 Use Scale II

18.010 ***Thought withdrawal***

Are your thoughts actually taken out or sent out of your mind? Do they actually feel like that? So that they are outside your head?

- *What is that like?*

R must describe active extraction, not 'thoughts seem to be outside my head'. Do not include 'as if' statements, e.g. my thoughts are so powerful that everyone must know them.

0 - 3 Use Scale II

18.011 ***Other subjective disorder of thought***

Is there any other kind of interference with your thoughts?

E.g. thoughts moved from left to right; cannot tell which are own thoughts etc.

0 - 3 Use Scale II

EXPERIENCE OF REPLACEMENT OF WILL

The essential element in the definition is that R experiences his/her will as replaced by the intentions of some other force or agency. The experience must be mood-incongruent. If it appears to be based on depression or elation it should not be rated present. A statement e.g. that the radio or some supernatural force is 'controlling' or 'influencing' the respondent without a description of the basic experience is insufficient. No positive rating should be made without unequivocal evidence and a recorded example. Do not make a positive rating if R is incapable of clear exposition.

If there is any possibility that involuntary possession experience (**2.057 - 2.059**) is present, consider also **19.024**.

18.012 ***Replacement of will by external force***

Do you feel that your will has been replaced by that of some force or power outside yourself?

- *Can you describe that?*
- *Is it like being a robot or zombie or puppet, controlled from elsewhere, without a will of your own?*
- *That your intentions have actually been replaced by those of ... ?*

Do not rate *'as if'* answers. The experience of replacement is essential. The will is experienced as diminished or replaced by that of some other agency. The expansion of will commonly experienced in elation, so that R feels, e.g. as powerful as if God were responsible (or that R is God), is not a delusion of control and should be excluded. Any answers obviously led by the questions must be verified against a free description by R. Experience of the phenomenon is essential.

0 - 3 Use Scale II

SKIP => to 18.018 if no evidence for experience of replaced will.

18.013 ***Replaced control of voice***

Does ... actually speak with your voice? You hear yourself saying things that you don't recognise and you didn't intend?

- *Does the voice seem to come from your own mouth?*

0 - 3 Use Scale II

18.014 ***Replaced control of handwriting***

What about your handwriting - do you seem to write things that you have not intended because it is under the control of ....?

0-3 Use Scale II

18.015 ***Replaced control of actions***

Do you actually seem to be a different person altogether, because your actions are outside your control?

- *Can you describe that?*
- *For example, were you made to walk, or run by ... ?*

0 - 3 Use Scale II

18.016 ***Replaced control of affect***

Are your emotions/feelings under the control of ... so that you do not recognise your emotions/feelings as your own?

0 - 3 Use Scale II

18.017 ***Other experiences of replaced control***

Is there any other kind of control, for example of your impulses? Or of your sensations?

Specify: be conservative in making this rating.

0 - 3 Use Scale II

18.018 ***Age at first onset of Section 18 symptoms***

Check dates at **1.013** and **1.017**.

18.019 ***Dates of PERIOD of Section 18 symptoms PE*** to

It is necessary to record only if dates of symptoms in this Section are different from PERIOD rated in Section 1 (items **1.007, 1.009, 1.010 -1.013**).

18.020 ***Interference with activities due to Section 18 symptoms***

You have mentioned [summarise symptoms] during PERIOD. Overall, how much interference has there been with your everyday activities because of these problems?

Rate interference due to symptoms in Section 20.

0 No Section 18 symptoms present in significant degree.
1 Symptom/s present but little interference.
2 Moderate or intermittent interference.
3 Severe to incapacitating interference.

Although opportunity to rate attribution of physical (including drug, medical condition, etc) cause is present using the optional attributional scale and the dashed boxes, item **18.021**and **18.022** allows the user to re-rate organic cause at the section / syndrome level.

Remember to complete course items for psychotic and affective disorders in Section 20 (**20.001 - 20.013**) if they are relevant.

Rate explanatory delusions at items **19.021 - 19.023**.

Rate insight at item **19.038**.

18.021 ***Organic cause of Section 18 symptoms***

0 Absent.
1 Probable organic cause; not fully confirmed.
2 Definite; confirmed by expert investigation.
8 Uncertain whether organic or not.

18.022 ***Identify the organic cause of Section 18 symptoms***

If alcohol or drugs (amphetamine, for example), are thought to cause Section 18 symptoms, rate in Sections 11 or 12, or/and with attributional scale at the item level. Other causes, e.g. Temporal Lobe Epilepsy, etc., enter ICD-10 chapter letter and up to 3 digits. If none leave blank.

19 Delusions

It will usually be evident by this stage in the interview whether delusions are present or not. Use judgement to how to word questions.

Note that rating Scale II is anchored on frequency or duration, which may not directly reflect on the clinical status of delusional beliefs.

19.001 ***Probe for delusions of reference***

Have you felt that people are unduly interested in you?
Or that things were arranged so as to have a special meaning.
Or even that harm might come to you?

- *Can you describe that?*

0 No.
1 Yes.

19.002 ***Probe for other delusions***

What about other unusual abilities or talents that some people have, such as having second sight, or being aware of strange powers or presences?

- *Are you superstitious?*
- *Do you have any special powers that most people lack?*
- *What is that like?*
- *Do you belong to a group of people who also have these experiences/powers?*

Include odd beliefs, magical powers, marked superstitions, clairvoyance, telepathy, etc. The experiences are not psychotic and may be normal within R's culture, religion or sect. However, they may also be indicators for individual psychopathology, in which case Sections 16 - 19 should be completed. Consider also dissociative symptoms rated at items **2.053 - 2.070**.

0 Absent.
1 Present.
2 Unsure.

CUT OFF => to 20.001 if no evidence of delusions.

DELUSIONS OF MISINTERPRETATION, MISIDENTIFICATION AND REFERENCE

19.003 ***Delusions of being spied upon***

Do people seem to talk about you, check up on you to find out where you are, or follow you about, or record your movements?

- *Do they take a special interest or try to photograph you?*
- *How do you know this?*

0 - 3 Use Scale II

19.004 ***Delusions of reference***

Do people seem to drop hints meant for you, or say things with double meanings?

- *Do you see messages for yourself in the newspapers or on TV or radio?*
- *Can you describe an example?*

0 - 3 Use Scale II

19.005 ***Delusional misinterpretation***

Do you sometimes see coded messages or a special significance in the way objects are arranged, or in colours, or in the way things happen?

- *Can you describe it?*

0 - 3 Use Scale II

19.006 ***Quotation of ideas***

Do you find that something you have previously thought or discussed is quoted on TV or in the newspapers or used in some other way to indicate a reference to you?

0 - 3 Use Scale II

19.007 ***Delusional misidentification***

Are there people about who are not what they seem? Who are perhaps in disguise?

- *Do you see people around whom you recognise from earlier in life?*
- *Can you give an example?*

0 - 3 Use Scale II

19.008 ***Familiar people impersonated***

Do you feel that the appearance of any people you know well has changed in ways that suggest that someone might be impersonating them?

This symptom may occur in a variety of clinical pictures. Rate all varieties of the delusion under this item.

0 - 3 Use Scale II

19.009 ***Delusional perception***

Take examples of delusions rated above. Remind R of the interpretation given.

When you [saw ... this event] how did you know what it meant?

- *Are you quite sure, or could you be mistaken?*
- *Is there no natural explanation?*
- *Have you had any previous experience that made you suspect something like this would happen?*
- *Did it come out of the blue?*

Intrusive, often sudden knowledge of a radically transformed meaning of a common perception. See definition for examples. Exclude if apparently based on abnormal mood, except delusional perplexity, or is part of a culture bound religious experience.

0 - 3 Use Scale II

19.010 ***Delusional ideas of reference based on guilt***

Must be understandable to some extent in terms of recent depressed mood, though the ideas may persist for a time after mood has improved: e.g. R has become depressed and guilty and now believes that when people shake their heads it means that she/he is to be executed for some nameless crime.

0 No such experience present.
1 Definite description of such an experience but frequency rare or difficult to rate.
2 Such experiences have occurred frequently.
8 Uncertain whether present, even after adequate examination.

19.011 ***Delusional ideas of reference based on expansive mood***

Must be understandable to some extent in terms of recent expansive mood, although the ideas may persist for a time after mood has improved: e.g. during a post-partum episode of elation and heightened sexuality, R thinks that the way doctors act mean that they are in love with him/her and are making him/her feel sexy.

NB: This is not a delusion of control.

0 No such experience present.
1 Definite description of such an experience but frequency rare or difficult to rate.
2 Such experiences occurred frequently.
8 Uncertain whether present, after adequate examination.

DELUSIONS OF PERSECUTION

19.012 ***Delusions of persecution***

Does anyone seem to be trying to harm you (trying to poison you or kill you?

- *Are they particularly singling you out?*
- *How do you experience this?*

Record description. Include explanations of other psychotic experiences in terms of persecution.

0 - 3 Use Scale II

19.013 ***Delusions of conspiracy***

Does there seem to be a plot or a conspiracy behind it?

- *How do you recognise it?*

Include explanations of other psychotic experiences in terms of conspiracy.

0 - 3 Use Scale II

Rate insight into delusions at **19.038**. Rate acting out delusions at **19.039**.

SEXUAL DELUSIONS

Sexual delusions based on hallucinations have been rated in Section 17. Rate here only delusions without evidence of hallucinations.

19.014 ***Delusional jealousy***

Do people say you are the jealous type?

- *Is it true? Are you jealous of your spouse/friend?*
- *What do you do to convince yourself that nothing is going on?*

Spies on actions. Smells clothes. Misinterprets, etc.

0 - 3 Use Scale II

19.015 ***Non-delusional jealousy (check at item 3.013.)***

0 - 3 Use Scale II

19.016 ***Delusions of pregnancy***

0 - 3 Use Scale II

19.017 ***Delusional lover***

Are you loved by someone who does not publicly acknowledge it?

- *Who is it?*
- *Was s/he the first to try to begin an affair?*
- *What evidence have you had of these advances?*
- *Do you try to make contact? In what way?*

An idealised love, usually with someone of higher status (de Clérambault syndrome) but not necessarily. May persistently follow and pester the supposed lover. Fully delusional. For sexual delusions associated with sexual hallucinations, see **17.027**.

0 - 3 Use Scale II

19.018 ***Delusion that others accuse R of homosexuality***

Do people seem to suggest that you are homosexual?

0 - 3 Use Scale II

OTHER DELUSIONS AND DELUSIONAL EXPLANATIONS

19.019 ***Delusional memories and fantastic delusions***

England's coast melting. Has lived for 100 years. Came down to earth on a silver star, is from the planet Pluto, etc.

0 - 3 Use Scale II

19.020 ***Preoccupation with previous delusions***

0 - 3 Use Scale II

19.021 ***Religious delusions***

What is the explanation for these experiences? Do you think there is a religious explanation?

Include only other psychotic experiences or delusional religious beliefs that are explanations of delusions and are themselves delusional. Be careful not to rate well accepted religious or spiritual beliefs, which may colour R's reaction to psychotic symptoms, as themselves delusional.

0 Not present.
1 Present.

19.022 ***Delusional paranormal explanations***

Is anything like hypnotism or telepathy going on?

- *What is it like?*

Occult influences, hypnotism, telepathy, ESP, etc.

0 Not present.
1 Present.

19.023 ***Delusional physical explanations***

Are you influenced or affected by X-rays, radio waves, neutrons, electrons, or machines or anything like that?

Radio waves, X-rays, laser beams, electricity, computers, television, machines etc.

0 Not present.
1 Present.

Rate insight into delusions at **19.038**. E.g. mental or physical illness. Unconscious thoughts, etc. Rate tendency to act out at **19.039**.

19.024 ***Specifically named local syndrome***

Annex 2 of ICD-10 DCR contains a proposed classification of Culture Specific Disorders. Most suggested ICD-10 codes are in F4, and occasionally F3 or F68.8. Examiners should be particularly cautious before making ratings in Part Two of SCAN on respondents who appear to exhibit features similar to those specified in Annex 2. The examiner should consider carefully ratings in Part One, Section 2 (dissociative and somatoform disorders), Sections 3 and 4 (anxiety disorders) and Section 6. The clinical history schedule (Section 27) should also be completed (i.e. **27.053** - **27.068**). Examiners who are not members of R's culture should carefully reconsider whether it is possible to complete, unaided, a reliable SCAN assessment (item **24.092**).

Include Koro, Latah, Witigo, or other such specified condition. The item should be rated 1 only if a full PSE10 interview has been completed, and ICD-10 DCR Annex 2 used to match clinical presentation with description in DCR.

0 Absent.
1 A full description of the state is obtained and PSE10 completed including Part One and Clinical History Schedule.

Specify the name of the state, and the main features shown: ....

DELUSIONS REVIEWED ELSEWHERE

These ratings will have been considered in the appropriate sections but should be reviewed here on the basis of the complete interview, particularly sections dealing with other psychotic symptoms. Also check delusions with hallucinations: **17.023** - **17.025**, **17.027** and **17.029**.

19.025 ***Delusions of guilt or worthlessness in context of depression***

Unshakeable conviction, in context of depression, of guilt, crime, evil, harm to others, worthlessness, etc., for which R feels culpable and deserving of punishment. Check **6.018**.

0 - 3 Use Scale II

19.026 ***Delusions of catastrophe in context of depression***

Unshakeable conviction, in context of depression that the world is about to end, the country is evil and will be destroyed, etc. R feels responsible. Check **6.019**.

0 - 3 Use Scale II

19.027 ***Hypochondriacal delusions in context of depression***

Unshakeable conviction that bowels are stopped up, insides are rotting, etc. in context of depressed mood. Check **6.020**.

0 - 3 Use Scale II

19.028 ***Hypochondriacal delusions without context of depression***

Rate delusional preoccupation with disease, which may involve the same symptoms as **19.027** but not the mood. Distinguish from delusions about appearance rated **16.012**; delusional explanations of somatic hallucinations rated at **17.029**. Check **2.038 - 2.039**.

0 - 3 Use Scale II

19.029 ***Delusions of grandiose abilities***

Check **10.016**.

0 - 3 Use Scale II

19.030 ***Delusions of grandiose identity***

Check **10.017**.

0 - 3 Use Scale II

19.031 ***Delusions concerning appearance***

Include dysmorphophobic delusions. Check **16.012**.

0 - 3 Use Scale II

19.032 ***Delusion of depersonalisation or realisation***

Delusion that some part of the body is missing; e.g. no head or no brain, no thoughts or no mind (symptom of Cotard). Some part of the external world may be missing. Include delusions of depersonalisation. Check **16.013**.

0 - 3 Use Scale II

GENERAL RATINGS

19.033 ***Duration of any delusion(s)***

1 Less than 1 week.
2 Less than 2 weeks.
3 Less than 1 month.
4 Less than 3 months.
5 Less than 6 months.
6 6 months or more.

19.034 ***Monothematic delusions***

Only one type of delusion is present though it may be predominant in the clinical picture and dominate the behaviour of R. Any development of delusions on other topics should be rated 1 here. Delusions that others think R smells, or is homosexual; delusions that R is pregnant, or has misshaped teeth; delusion of jealousy, etc. Direct elaboration of the central delusion e.g. when a jealous R interprets a light being switched on as a signal to a lover, is acceptable.

0 No delusions.
1 Delusions are not monothematic.
2 Monothematic delusion (plus accessories) only.

19.035 ***Systematisation of delusions***

Base the rating on the extent to which all the delusions have a common theme or development.

0 No delusions.
1 No systematisation present. Delusions unrelated.
2 Some systematisation.
3 Close systematisation. Most delusions are related to one 'scenario' though this may undergo elaborations or modifications from time to time.
4 Completely systematized delusions.

19.036 ***Prominence of delusions***

0 No delusions.
1 Delusions present but not a central feature of the clinical picture.
2 Delusions are a prominent feature of the clinical picture compared to other symptoms.

19.037 ***Mood-incongruent delusions***

Take into account all delusions present.

0 No delusions.
1 Most delusions are congruent with mood.
2 Mixed congruent and incongruent.
3 Most delusions are mood-incongruent.

19.038 ***Conviction about delusions or hallucinations [insight]***

Include all symptoms in Sections 17 - 19 rated positively. Consider explanations in terms of natural causes as well as those rated at items **19.021 - 19.024**.

0 No delusions or hallucinations.
1 Aware of delusions or hallucinations during period but also aware of their abnormal delusional nature.
2 Brief periods of doubt but generally convinced.
3 Unshakeably convinced.

19.039 ***Actions based on delusions and/or hallucinations***

0 No delusions or no acting out.
1 Actions without serious social or family effects.
2 Some seriously embarrassing or distressing effect.
3 Aggressive or violent consequences.

19.040 ***Bizarreness of delusions***

Rate bizarreness of delusions according to the definition given in the Glossary.

0 Absent.
1 Present.

19.041 ***Age at first ever onset of delusions***

Check **1.017**.

19.042 ***Dates of PERIOD of Section 19 symptoms*** to

It is necessary to record only if dates of symptoms in this Section are different from PERIOD rated in Section 1 (items **1.007, 1.009, 1.010 -1.013**).

19.043 ***Interference with activities due to Section 19 symptoms***

You have mentioned [summarise symptoms] during PERIOD. Overall, how much interference has there been with your everyday activities because of these problems?

Rate interference due to symptoms in Section 20.

0 No Section 19 symptoms present in significant degree.
1 Symptom/s present but little interference.
2 Moderate or intermittent interference.
3 Severe to incapacitating interference.

Although opportunity to rate attribution of physical (including drug, substance etc) cause is present using the optional attributional scale and the dashed boxes, items **19.044**and **19.045** allows the user to re-rate organic cause at the section / syndrome level.

19.044 ***Organic cause of Section 19 symptoms***

0 Absent.
1 Probable organic cause; not fully confirmed.
2 Definite; confirmed by expert investigation.
8 Uncertain whether organic or not.

19.045 ***Identify the organic cause of Section 19 symptoms***

If alcohol or drugs (amphetamine, for example), are thought to cause Section 19 symptoms, rate in Sections 11 or 12. Other causes, e.g. Temporal Lobe Epilepsy, etc., enter ICD-10 chapter letter and up to 3 digits. If none, leave blank.

Remember to complete course items for psychotic and affective disorders [**20.001 - 20.013**] if relevant.

20 Further information for classification of Part Two disorders

[It is only necessary to code the history items once. Use the present state period schedule if there are two periods.]

It is essential, particularly for users of DSM, to rate **20.001 - 20.092**, Section 23 (Observed affect), Section 24 (Speech), Section 25 (Social impairment) and Section 27 (Role performance items) if there is any possibility that the diagnosis is schizophrenia or delusional or affective psychosis. The actual classification will be made by the computer program. Slight differences between items are due to rules for different systems.

20.001 ***Length of episode of schizophrenia***

Enter the length in months of the longest episode of schizophrenia including prodromal, active and residual phases.

888 = NK.

NB: Criteria for the active phase are summarised in **20.002**. The prodromal and residual phases are characterised by negative symptoms (items **20.76 - 20.91** and Section 22 - 24, noting in particular items, **22.011**, **22.012**, **22.013**, **22.040**, **22.050**, **23.008**, **24.016**, **24.019**, **24.020**, **24.027**) and/or odd beliefs (e.g. **14.006**, **20.033**) and behaviour not amounting to fully florid symptoms (e.g. **14.012**, **20.030 - 20.042**, **22.035 - 22.037**, **22.047**, **22.048**). Rate deterioration in functioning and features of the prodomal and residual phases in sections 22 - 24 (negative features) and the clinical history schedule (deterioration in social and occupational functioning, N.B. rate cause of Social handicap, item **27.051**). These ratings are required for DSM schizophrenic disorders.

20.002 ***Duration of active symptoms included*** ☐

If **20.001** is rated 6+, did the episode include at least 1 to 4 weeks of active symptoms? These include at least one of the following: Section 18 symptoms, "bizarre" delusions, voices commenting on or discussing R, or two of the following: prominent hallucinations, incoherence, catatonic behaviour, and flat or inappropriate affect.

0 The episode did not include a week of such symptoms.
1 Less than 1 week because of succesful treatment.
2 The episode did include 1 week of such symptoms, but less than 1 month.
3 Less than 1 month because of succesful treatment.
4 The episode did include at least 1 month of such symptoms.
5 No prominent mood symptoms.
6 Extensive period of active phase symptoms not amounting to much of the total duration of **20.001**.
7 Active phase symptoms only, through **20.001**.
8 Uncertain.
9 NA e.g. **20.001** is rated < 6.

The active phase may be shortened by rapidly effective treatment but there must be no doubt active symptoms were clearly present when treatment was commenced. Do not rely only on informants. If there genuine doubt, rate 8 (uncertain). If no negative features were observed (differentiate from side effects of drug treatment) and any deterioration in functioning began with active phase symtoms and resolved rapidly when active phase symptoms subsided, then consider ratings 6 or 8.

20.003 ***Course of schizophrenia DSM*** ☐

The course of schizophrenia is rated as a fifth digit.

0 Not rated.
1 **Subchronic:** Prodromal, active and residual phases of the episode lasted more than 6 months but less than 2 years.
2 **Chronic:** As (1) but for more than 2 years.
3 **Subchronic with acute exacerbations:** As (1) with re-emergence of active florid symptoms during residual phase.
4 **Chronic with acute exacerbations:** As (3) but re-emergence during residual phase of chronic disorder.
5 **In remission:** Past history of schizophrenia but no symptoms of the disorder for at least 3 months before the key date. R may still be on medication.

20.004 ***Duration of affective/schizophrenic episodes***

If full manic or depressive syndromes were present during the clinical course as well as the positive schizophrenia syndrome: was the total duration of all affective episodes brief compared with the total length of the active and residual phases of schizophrenia?

0 Not rated.
1 Duration of schizophrenia longer.
2 Manic/Bipolar episodes longer.
3 Major Depressive episodes (no mania) longer.
8 Unclear which was predominant.
9 NA.

NB: This item is concerned with the full affective and positive schizophrenic syndromes, not with isolated symptoms. Rate on the whole course. If uncertain, rate 8.

20.005 ***Autistic syndrome present as well as schizophrenia***

Are prominent hallucinations and or delusions present, in addition to Pervasive Developmental Disorders (Autism, Asperger's syndrome: Section 25)?

0 Not rated.
1 Autistic syndrome present without prominent hallucinations or delusions.
2 Autistic syndrome present and there are also prominent hallucinations or delusions.
8 NK.

Items **20.006 - 20.012** refer chiefly to ICD-10, but some also are relevant to DSM.

20.006 ***Affective syndromes present with delusional syndrome***

Was the total length of all affective syndromes brief compared with the delusional disturbance?

0 Not rated.
1 Delusional syndrome relatively longer.
2 Affective syndrome relatively longer. Chiefly manic or bipolar.
3 As (2). Chiefly depressive.
4 Affective and delusional paranoid syndrome both prominent and concurrent.
8 Unclear whether affective or delusional syndrome was longer.

NB: This item is concerned with the full affective and delusional syndromes, not with isolated symptoms. Rate on the whole course. If uncertain, rate 8.

20.007 ***Length of episode of schizophrenia***

If the general criteria for schizophrenia are fulfilled did the episode last more or less than one month?

0 Not rated.
1 Episode lasted less than 1 month.
2 Episode lasted at least 1 month.

20.008 ***Course of schizophrenia ICD-10***

The course of schizophrenia is rated as a fifth character if there has been at least one year's course.

0 Continuous. No remission of psychotic symptoms throughout the period of observation.
1 Episodic with progressive development of negative symptoms in the intervals between psychotic episodes.
2 Episodic with persistent but non-progressive negative symptoms in the intervals between psychotic episodes.
3 Episodic with complete or virtually complete remissions between psychotic episodes.
4 Single psychotic episode followed by an incomplete remission.
5 Single psychotic episode followed by a complete or virtually complete remission.
6 Any other pattern.
8 Course not assessable.
9 NA. e.g. less than one year's course.

20.009 ***Previous positive schizophrenia***

If positive schizophrenia is absent at the time of key examination, has a previous positive episode, lasting at least one month occurred?

0 Not rated.
1 No previous episode of positive schizophrenia.
2 At least 1 previous episode of positive Schizophrenia.
8 NK.
9 NA.

20.010 ***Relation of affective syndromes to schizophrenic syndrome***

Did positive schizophrenic symptoms begin before onset of affective symptoms? Which syndrome predominated?

0 Not rated.
1 Affective syndrome first.
2 Affective syndrome dominate.
3 Affective syndrome first and dominate.
4 Schizophrenic syndrome first.
5 Schizophrenic syndrome dominate.
6 Schizophrenic syndrome first and dominate.
8 Uncertain.
9 NA.

NB: This item is concerned with the full affective and schizophrenic syndromes, not with isolated symptoms. If the schizophrenic syndrome developed first and/or clearly dominated the clinical picture, rate 2, even if a full manic or depressive syndrome was also present during the period. If uncertain, rate 8.

20.011 ***Post-schizophrenic depression***

A depressive episode beginning within a year of a schizophrenic episode; at least 1 schizophrenic symptom still present. Episode/s of depression must be recorded in Section 6.

0 Not rated.
1 Not post-schizophrenic depression.
2 Probably PSD.
3 Probably PSD (and schizophrenia items full rated).
8 Uncertain.
9 NA.

20.012 ***Duration of delusional syndrome***

How long have 'non-schizophrenic' delusions (reference, persecution, expansive, depressive, monosymptomatic, etc.) persisted?

Enter number of months.

888 = NK.
999 = NA.

20.013 ***Delusions or hallucinations unchanged by affective syndrome***

If delusions or hallucinations have been co-terminous with affective symptoms at some time, have they also persisted unchanged in the absence of the affective symptoms?

0 Not rated.
1 Delusions or hallucinations have always been co-terminous with affective symptoms except for very brief periods (1 or 2 days).
2 Delusions or hallucinations have persisted unchanged in the absence of affective symptoms for at least one period lasting at least two weeks
3 Delusions or hallucinations have persisted unchanged in the absence of affective symptoms for at least one period lasting more than a month or longer.
9 NA.

Rate factors influencing the manifestation of Section 16 - 19 symptoms at items **20.045 - 20.060**.

OPTIONAL CHECKLIST (items 20.014 - 20.060)

SKIP => to 20.061 if optional checklists not required or applicable.

ACUTE PSYCHOSIS

Items **20.014 - 20.025**, together with other items rated elsewhere, are needed for several diagnostic symptoms. If the systems follow acute stress, rate items **13.064 - 13.066** as well.

20.014 ***Acute onset with delusions, hallucinations, thought disorder or incoherence in any combination, lasting at least several hours.***

0 Absent.
1 Present.

20.015 ***Full development of psychotic state.***

1 Within 48 hours.
2 Within 2 weeks.
3 More than 2 weeks.

20.016 ***Duration, onset to recovery, treated or not.***

1 up to 1 month.
2 1 - 3 months.
3 3 months +.

20.017 ***Poorly organised, polymorphic symptoms, shifting rapidly changing from day to day in type and intensity.***

0 Absent.
1 Present.

20.018 ***A major precipitating stress within 2 weeks of onset.***

0 Absent.
1 Present.

20.019 ***Emotional turmoil, ecstatic, anxious, other affects by turns.***

0 Absent.
1 Present.

20.020 ***Affective syndrome*** (This must be rated in Part One of SCAN).

0 Absent.
1 Present.

20.021 ***Perplexity or misidentification of people or places.***

0 Absent.
1 Present.

20.022 ***Marked increase or decrease in motility.***

0 Absent.
1 Present.

20.023 ***Age at onset of Acute Psychosis.***

Although opportunity to rate attribution of physical (including drug, medical condition, etc) cause is present using the optional attributional scale and the dashed boxes, items **20.024** and **20.025** allow the user to re-rate organic cause at the section/syndrome level.

20.024 ***Organic cause of Acute Psychosis.***

20.025 ***Identify organic cause of Acute Psychosis.***

Enter ICD-10 code:

INDUCED PSYCHOTIC DISORDER

Includes 'Folie a deux' and also induced psychosis in members of a larger social group, several of whom meet the criteria. All 3 symptoms must be present and have developed over at least a year. The delusions must meet the criteria listed in Section 19 of the Glossary. Consider dissociative symptoms, items **2.049 - 2.065 and 17.020**.

Rate items **20.026 - 20.028**.

0 Absent.
1 Present.

20.026 ***Close relationship***

Two or more people support each other in a close relationship, relatively isolated from others.

20.027 ***Induced delusions***

The delusions in R, the passive member, are induced by the active member (F20 - 23).

20.028 ***Previous psychotic state***

The passive member had delusions before meeting the active, and was psychotic (F20 - 23).

20.029 ***Age at onset of Induced Psychosis***

NK = 88.

SCHIZOTYPAL DISORDER (latent schizophrenia)

In DSM-III-R, this is a personality disorder, as is Schizoid Personality and Paranoid Personality. Personality Disorders are not rated in PSE10.

Schizotypal Disorder in ICD-10 is category F21.0, the items in which are listed below. The condition must have lasted at least 2 years, with several of the symptoms being present continuously or repeatedly.

Rate items **20.030 - 20.038**:

0 Absent.
1 Present.

20.030 ***Cold and aloof, inappropriate or few strong emotions.***

20.031 ***Odd, eccentric, peculiar appearance or behaviour.***

20.032 ***Poor rapport, tending to social withdrawal.***

20.033 ***Odd or bizarre beliefs, or magical thinking influencing behaviour and inconsistent with subcultural norms, but not true delusions.***

20.034 ***Suspiciousness or paranoid ideas.***

20.035 ***Ruminations without inner resistance. Often sexual, aggressive or dysmorphophobic contents.***

20.036 ***Unusual perceptual experiences including somatosensory (bodily) or other illusions, depersonalisation or derealisation occasionally.***

20.037 ***Vague, circumstantial, metaphorical, over-elaborate speech but not gross incoherence or irrelevance.***

20.038 ***Occasional transient quasi-psychotic episodes: illusions, hallucinations, delusional notions, without external provocation.***

20.039 ***Age at onset of Schizotypal disorder.***

SIMPLE SCHIZOPHRENIA

This category does not occur in DSM-III-R. The following criteria are for ICD-10. If some or all are met complete Section 25. There must have been a slow progressive development, over at least 1 year, of all the following elements.

Rate items **20.040 - 20.043**.

0 Absent.
1 Present.

20.040 ***Significant and consistent change in overall quality of some aspects of personality behaviour: loss of drive and interests, becoming idle and aimless, self-absorbed and socially withdrawn.***

20.041 ***Gradual appearance and deepening of negative symptoms: underactive, blunted, lack of initiative, poverty of content and quantity of speech, poor non-verbal communication.***

20.042 ***Decline in social, scholastic or occupational performance.***

20.043 ***Previously positive psychotic symptoms.***

20.044 ***Age at onset of Simple Schizophrenia.***

Interference with activities due to Part Two symptoms and factors influencing their manifestation (attributions).

If two periods have been rated in Sections 16 - 20, attributional ratings here apply to the left hand box records, i.e. present state PERIOD.

SKIP => to Section 13 **if the main problems are related to Part One symptoms.**

SKIP => to Section 21 **if there are no or only minimal Section 16 - 26 problems.**

Major cause of interference with activities due to Part Two symptoms

Review ratings of interference due to syndromes in previous Sections. Consider whether any one type of symptom or syndrome causes the greatest interference. Check that the corresponding interference rating is higher than that of other similar ratings. If necessary ask:

You have mentioned [summarise symptoms] *during* [PERIOD]. *Overall, how much interference has there been with your everyday activities because of these problems?*

Relationship of Part Two symptoms and syndromes to each other:

Reconsider relationship between Sections in which items have been rated positively. Reconsider if necessary ratings of items **17.014** and **19.041**. If not clear which symptoms or syndromes are primary, consider asking:

You have mentioned [summarise symptoms] *during* [PERIOD].*Can you decide which of these types of symptoms or complaints is the most important for you?*

Does any one type of problem (prompt with examples of experiences, beliefs etc.) seem to be the main problem that the others follow from?

If yes:

Could you say, if possible, which is next in importance?

Factors influencing manifestation of Sections 16 - 26 symptoms:

Both ICD-10 and DSM-III, III-R and IV, require the exclusion of attributions of "cause", particularly in relation to the judged effects of psychoactive substances and alcohol, other treatments, physical (or "organic") disease, and psychosocial stressors and adversity. Users of SCAN who wish to make use of such classification systems must therefore attempt to consider the grounds for making judgements of this kind, here in relation to Part Two symptoms. Although opportunity to do so at the item level using the etiology option and scale is present in earlier sections, Section 20 allows the user to re-rate these at a syndromal level, in order to permit the scientific investigation of the reliability and validity of such attributions.

Note: Users should establish that ratings of attribution can be made *reliably*. As far as is practicable, such attributional ratings should be delayed until the "causal" status of any attribution can be established with reasonable certainty.

Criteria for rating Attributions of Cause:

(Taken from ICD-10 DCR).

If the following three criteria are met, a *provisional* attribution (i.e. rate 1) can be made:

1. There is objective evidence (from physical and neurological examination and laboratory tests) and/or history of cerebral disease, damage, or dysfunction, or of systematic physical disorder known to cause cerebral dysfunction, including hormonal disturbances (other than alcohol- or other psychoactive substance-related) and non-psychoactive drug effects.

2. There is a presumed relationship between the development (or marked exacerbation) of the underlying disease, damage, or dysfunction, and the mental disorder, the symptoms of which may have immediate onset or may be delayed.

3. There is insufficient evidence for an alternative causation of the mental disorder, e.g. strong family history of a clinically similar or related disorder.

Rate 1, if the above three criteria are satisfied. If not rate 0. If, in addition the following criterion can be met, the attribution can be regarded as *more certain* and rate 2. If in doubt, rate 8. Keep careful records of the evidence upon which ratings are based:

4. There is recovery from or significant improvement in the mental disorder following removal or improvement of the underlying presumed cause.

20.045 ***S16. Perceptual disorders other than hallucinations.*** ☐

20.046 ***S17. Auditory hallucinations (17.003-17.013).*** ☐

20.047 ***S17. Visual hallucinations (17.015-17.019).*** ☐

20.048 ***S17. Other hallucinations (17.022-17.030).*** ☐

20.049 ***S18. Subjectively described thought disorder (18.001, 18.003-18.011).*** ☐

20.050 ***S18. Experience of replacement of will (18.012-18.017).*** ☐

20.051 ***S19. Delusions (19.003-19.040).*** ☐

20.052 ***S21. Cognitive impairment/decline.*** ☐

20.053 ***S22. Motor, movement and behavioural disorders.*** ☐

20.054 ***S23. Affect.*** ☐

20.055 ***S24. Speech.*** ☐

20.056 ***S25. Social impairment.*** ☐

20.057 ***S27. Problems recorded only in Clinical History Schedule.*** ☐

20.058 ***Other specified................................*** ☐

20.059 ***Unknown or multiple Part Two symptoms.*** ☐

20.060 ***Quality of data available to rate physical/drug/substance/ organic attribution of cause:***

0 No such attribution, Part Two of SCAN.
1 Attribution suspected but inadequate data available.
2 Confident attribution can be based on clinical data.
3 Confident attribution can be based on clinical and laboratory data.

SKIP => to 20.076 if 20.045 - 20.060 all rated 0.

Rate here attributions of the effects of psychosocial, substance use or physical factors. It is assumed that these ratings apply to the left hand box ratings or to both the left and right.

20.061 ***Identify first attribution (i.e.) "organic cause" of Part Two symptoms.***

For ratings in items **20.045 - 20.060** enter section number (i.e. 16 - 27) and ICD-10 Chapter identifier and 1-3 digits. See Appendix to SCAN Glossary for a list of conditions that may be associated. Leave blank if no attribution has been made in **20.045 - 20.060** or with the optional attribution scale. Include dependence-related substances such as alcohol and opioids, etc. These should also be rated as causes of abnormal mental states in Sections 11 and 12. Leave blank if none.

20.062 ***Re-rate interference if altered by this attributed cause.***

0, 1, 2, 3

20.063 ***Identify second attribution of cause of Part Two symptoms, rated in 20.045 - 20.060.***

If a second attribution of cause is required Enter ICD-10 Chapter identifier and 1-3 digits. Leave blank if none. Include dependence-related substances such as alcohol and opioids, etc. These are also rated as causes of abnormal mental states in Sections 11 and 12. Leave blank if none.

20.064 ***Re-rate interference if altered by attributed cause.***

0, 1, 2, 3

20.065 ***Identify third attribution of Part Two symptoms.***

If a third attribution of cause is required Enter ICD-10 Chapter identifier and 1-3 digits. Leave blank if none. Include dependence-related substances such as alcohol and opioids, etc. These are also rated as causes of abnormal mental states in Sections 11 and 12. Leave blank if none.

20.066 ***Re-rate interference if altered by third attributed cause.***

0, 1, 2, 3

Psychosocial attributions influencing manifestation of Sections 16- 26 symptoms

The principles underlying the above criteria for rating the attribution of physical cause, appropriately translated, may also be applied to attributions of the effects of therapy, including psychosocial treatment, and the effects of psychosocial stressors. The attribution of abnormal traumas and stressors is required for ICD-10 and DSM (all revisions since III) and is dealt with in Section 13 (stress and adjustment disorders). Stress induced psychoses as defined in ICD-10 can also be rated in this Section, as can the attribution of psychosocial cause, treatment etc.

Items/symptoms should have been rated already in the appropriate sections of Part Two (item **20.018, 20.026 - 20.028**). These attributions are only required for symptoms that are definitely present during the period.

Did adverse events or circumstance have any influence or cause the problems you have described (prompt Part One symptoms) during the PERIOD?

- *How important were they?*
- *For example:* [use prompt card if it would be helpful.]
- *Were they the main cause or did they just make things worse or keep them going?*

0 Factor absent.
1 Factor present but effect uncertain.
2 Factor present and modifies the clinical picture or course. Not the main cause or precipitant.
3 Factor the main cause or precipitant and/or markedly modifies clinical picture or course.

20.067 ***Childbirth (only rate 2 or 3 if onset within 28 days), Part Two*** ☐

20.068 ***Menopause, Part Two*** ☐

20.069 ***Bereavement (only rate 2 or 3 if onset within 3 months), Part Two*** ☐

20.070 ***Adolescence, Part Two*** ☐

20.071 ***Member of a special cultural group, Part Two*** ☐

20.072 ***Recent migration, Part Two*** ☐

20.073 ***Life events (use special schedule), Part Two*** ☐

20.074 ***Social deprivation, Part Two*** ☐

20.075 ***Other, specify, Part Two ....*** ☐

NEGATIVE SYNDROME AND ITEMS

The negative syndrome of schizophrenia ("deficit syndrome") is an optional subsection which is included for research purposes. Items **20.088 - 20.091** are present to allow ICD-10 diagnosis of related subtypes of schizophrenia.

The ratings are based on the presence of negative symptoms but must also include a clinical judgement as to whether a symptom is primary or secondary, and ascertainment of the duration of the symptom. The symptom must have been present for the preceding 12 months and was always present during periods of clinical stability (including chronic psychotic states). The symptom may or may not have been detectable during transient episodes of acute psychotic disorganization or decompensation.

Symptom items have similar rating points with specific criteria. Etiologic attribution for each negative symptom item is rated in the following item with a 6-point rating scale which allows atribution of more than one factor. The attribution scale is:

Attribution Scale for Negative Syndrome

0 Primary symptom, not secondary to other factors.
1 Due to anxiety.
2 Due to drug effects.
4 Due to suspiciousness (and other psychotic symptoms).
8 Due to mental retardation.
16 Due to depression.

To score more than one of these just circle the items and add up the numbers for a total score to be recorded. This number can always be uniquely decomposed into the original constituents for computer scoring. **This scale is not the general attributional scale and is to be used only in this subsection.**

20.076 ***Restricted affect***

This item is to be rated on the basis of what is observed in the interview and by others who have longterm contact with the subject. Restricted affect refers to observed behaviors rather than the subject's subjective experience. Specifically, one rates:

a. A relatively expressionless face, or unchanging facial expression.
b. Reduced expressive gestures when emotional material is discussed.
c. Diminished vocal inflection.

It is important to distinguish primary restricted affect from a guarded speaking style which is caused by suspiciousness, or a relatively normally reticence or shyness in an interview. These secondary features should be distinguished from a truly restricted affect, and should lead to a rating of 0 or 1 on this item. Information from other sources may be helpful in making such distinctions.

0 The subject has a spontaneous and widely ranging display of affect.
1 A decrease in affect displayed so mild it is not confidently considered pathological.
2 A moderate decrease of affect displayed in relation to many topics, or a severe decrease in some topics.
3 A severe decrease in affect displayed in relation to many topics, or a very severe decrease in relation to some topics.
4 Symptom present but not for the preceding 12 months.

20.077 ***Restricted affect - primary or secondary***

(Use attribution scale for Negative Syndrome)

20.078 ***Diminished emotional range***

What do you really enjoy about life?
Do other people seem to enjoy things more than you do?
What upsets you?
What makes you sad?
What irritates you or makes you angry?

0 No diminution of emotional range or intensity.
1 A questionable decrease in variety or intensity of emotions of such mild severity that it is not confidently considered pathological.
2 A moderate decrease in variety or intensity of emotions in relation to many activities, or a severe decrease in relation to some activities.
3 A severe decrease in variety or intensity of emotions in relation to many activities, or a very severe decrease in relation to some activities.
4 Symptom present but not for the preceding 12 months.

20.079 ***Diminished emotional range - primary or secondary***

(Use attribution scale for Negative Syndrome)

20.080 ***Poverty of speech***

0 Speech normal in quantity and amount of information conveyed.
1 A decrease in the quantity of speech and the amount of information conveyed so mild that it is not confidently considered pathological.
2 A moderate decrease in the quantity of speech and the amount of information conveyed in relation to many topics, or a severe decrease in relation to some topics.
3 A severe decrease in the quantity of speech and the amount of information conveyed in relation to many topics, or a very severe decrease in relation to some topics.
4 Symptom present but not for the preceding 12 months.

20.081 ***Poverty of speech - primary or secondary***

(Use attribution scale for Negative Syndrome)

20.082 ***Curbing of interests***

What do you find interesting? What do you spend a lot of time thinking about?
What do you do to find out about these things?
What would you like to know more about?
Have you read the newspapers, or listened to the news on TV or radio recently?
Can you tell about a big news story you have heard about recently?

0 A normal amount of interest in the world: either a wide range of interests or intense involvement in a narrow range of interests.
1 A decrease in the range or intensity of interests so mild that it is not confidently considered pathological.
2 A moderate decrease in the range or intensity of interests.
3 A severe decrease in the range or intensity of interests.
4 Symptom present but not for the preceding 12 months.

20.083 ***Curbing of interests - primary or secondary***

(Use attribution scale for Negative Syndrome)

20.084 ***Diminished sense of purpose***

0 Normal motivation and goal-directed activity displayed either in a relatively wide range of areas, or intensely in a narrow range.
1 A decrease in motivation and goal-directed activity that is so mild that it is not confidently considered pathological.
2 A moderate decrease in range or intensity of motivation and goal-directed activity
3 A severe decrease in range or intensity of motivation and goal-directed activity.
4 Symptom present but not for the preceding 12 months.

20.085 ***Diminished sense of purpose - primary or secondary***

(Use attribution scale for Negative Syndrome)

20.086 ***Diminished social drive***

Who do you spend time with?
Do you usually do things alone or with other people?
Some people like to be with others; some people prefer to be alone. How would you describe yourself? (if the subject is withdrawn) Why do you prefer to be alone?
Do you get lonely sometimes? Do you want to spend more time with people?
Do you seek people out?
Do you usually wait for others to ask you to do something with them?
When you get together with people, who decides what to do?
When you spend time with others, do you initiate it, or do they?

0 Spontaneous expression of social drive in a relatively wide range of situations.
1 A decrease of social drive that is so mild that it is not confidently considered pathological.
2 A moderate decrease of social drive.
3 A severe decrease of social drive.
4 Symptom present but not for the preceding 12 months.

20.087 ***Diminished social drive - primary or secondary***

(Use attribution scale for Negative Syndrome)

20.088 ***Poor social performance***

Do you have trouble interacting with other people?
Do other people make fun of you or tease you?
How confident do you feel dealing with other people?

0 No problems with social performance.
1 A decrease of social performance that is so mild that it is not confidently considered pathological.
2 A moderate decrease of social performance.
3 A severe decrease of social performance.
4 Symptom present but not for the preceding 12 months.

20.089 ***Poor self care***

0 Normal self care.
1 A decrease of self care that is so mild that it is not confidently considered pathological.
2 A moderate decrease of self care.
3 A severe decrease of self care.
4 Symptom present but not for the preceding 12 months.

20.090 ***Poor non-verbal communication***

0 Normal non-verbal communication.
1 A decrease of non-verbal communication that is so mild that it is not confidently considered pathological.
2 A moderate decrease of non-verbal communication.
3 A severe decrease of non-verbal communication.
4 Symptom present but not for the preceding 12 months.

20.091 ***Psychomotor slowing or underactivity***

0 Normal psychomotor activity.
1 A decrease of psychomotor activity that is so mild that it is not confidently considered pathological.
2 A moderate decrease of psychomotor activity.
4 Symptom present but not for the preceding 12 months.

ADEQUACY OF INTERVIEW

20.092 ***Adequacy of ratings in sections 16 - 20***

0 No difficulties in making ratings.
1 Some problems but most ratings reasonably accurate.
2 Major difficulty in conducting examination. Ratings must be interpreted with caution.
3 Impossible to rate these sections with any confidence.

Remember to check number and dates of episodes and first onsets in Section 1 and if different within each Section on the basis of information provided during the interview.

21 Cognitive impairment and/or decline

Section 21 has two different functions:

(a) to guide a clinical assessment of level and pattern of cognitive impairment - for example in schizophrenia, depression, mild brain damage, mental retardation, Asperger's syndrome or other developmental disorders. Some questions also help to assess educational skills.

(b) to measure degree and type of cognitive decline, usually in the elderly. The chief conditions involved are dementia, amnestic syndromes and delirium. Associated causes or pathologies are rated at **21.103 - 21.124**.

It is therefore essential to rate the whole of the section, including the behavioural subsection (**21.065 - 21.076**) and the subsections based on information from records, tests or informants (**21.077 - 21.088**). An associated instrument with neuropsychological tests is in development.

The face sheet (Section 00) plus performance during the PSE will give relevant information concerning cognitive functioning.

21.001 ***Evidence of cognitive decline or impairment.*** ☐

0 Specific cognitive screening not performed, but information from the remainder of the examination is sufficiently detailed to presume no gross evidence of cogntive impairment or decline.
1 R is unable to participate in test or interview questions because of cognitive impairment from any cause.

Rate 2 in box and **SKIP => to 21.065.**

2 Proceed to **21.002**.
9 Not given because of time constraints or for other reasons, but not 1.

Always administer **21.002 - 21.028** if at all possible.

ADMINISTER VERBAL TRAILS TEST

Make sure R can count to 10 and say/recite alphabet.

21.002 ***Verbal trails test***

I am going to ask you to mix the alphabet with the numbers:
1 - A, 2 - B, and keep going.....

Time the task allowing up to 2 minutes. Stop after R reaches "13 - M". If an error occurs, e.g. "1 - A, 2 - B, 2 - C", correct it immediately, e.g. "3 - C". Use the rating scale to record the time of the task. Record the number of errors separately.

0 0-34 sec.
1 35-44 sec.
2 45-54 sec.
3 55-64 sec.
4 65-74 sec.
5 75-89 sec.
6 90-104 sec.
7 105-119 sec.
8 2+ minutes, unable to complete task due to cognitive impairment.
9 Not attempted for reasons othe than cognitive impairment.

ADMINISTER MINI-MENTAL STATE

Always administer all the test items if at all possible.

If there are motor problems and/or impairments of willed action associated with brain disorders such as Parkinsonism - slow, perhaps jerky, movements and speech, with reduced motivation to participate in tests, so that neither 0 nor 1 is a suitable rating - rate 6. Insert diagnosis at **21.103 - 21.125**.

Score items in MMS:

0 Incorrect.
1 Correct.
6 Motor/motivational disorder may explain difficulty in answering (see above).
7 Unable to attempt due to lack of education or unfamilitary with local geography.
8 Unable to attempt for physical reasons, e.g. blindness etc.
9 Unable to attempt answer because of cognitive impairment.

Always continue if any suspicion of cognitive disorder.

Maximum MMS item score at side of box.

21.003 *What day of the week is it?*

21.004 *What is the date today?*

21.005 *And the month?*

21.006 *And what year is it?*

21.007 *What is the season of the year?*

21.008 *Can you tell me the name of this County [Suburb/District]?*

21.009 *What is the name of this town [city/village]?*

21.010 *And what is the name of the street?*

- *(A street near here?)*

21.011 *And the name of this building?*

- *(Name and number of this house?)*

21.012 *What country are we in?*

21.013 *Now I am going to tell you the names of three objects. When I have finished I want you to repeat and remember them - apple, table, penny. Could you repeat these names please?*

Score number correct. Maximum 3.

21.014 Number of trials (registration)

If any incorrect repeat up to 6 trials.

0 Correct first time.
1-4 Extra trials.
5 Criterion not reached in 5 trials.

Do not count in total MMS score.

Note the exact time. R will be asked to recall names of these 3 objects twice; Once after attempting to subtract serial 7s and or/ spell "world" backwards and again after exactly 5 minutes.

21.015 *Now I would like you to take 7 away from 100 then go on counting backwards, taking 7 off each time.*

Stop after 5 subtractions: 93, 86, 79, 72, 65.

Score number correct. Maximum 5. Count only one error if R makes a subtraction error, but subsequent answers are 7 less than the error.

Score calculation/attention. Maximum 5.

21.016 *Could you spell 'world' backwards please?*

Spell 'world' forwards for R. Score number of letters in correct order, e.g. 'dlrow' = 5, 'dlorw' = 3.

* Maximum MMS score for **21.015** and **21.016** is 5 (Count highest score of these 2 items when manually summing total MMS score).

21.017 *Now could you tell me the three objects that I named a few minutes ago?*

Score recall. Maximum 3.

21.018 [show pencil] *What is this?*

21.019 [show watch] *What is this?*

21.020 *Please repeat after me: 'No ifs, ands or buts.'*

One trial only.

21.021 [grasp paper with both hands.]

Take this paper in your right hand, fold it in half, and put it on the floor. [Give paper extending both hands].

Score 1 point for each correct action. Maximum 3.

21.022 [show paper] *'Close your eyes'*

Please read this sentence aloud and do what it says.

0 Does not read sentence.
1 Reads and closes eyes.

SKIP => to 21.026 If R has followed the instructions or has correctly read the sentence.

If R neither follows instructions nor reads the sentences aloud:

21.023 [show paper again]

Can you read this sentence aloud?

0 R correctly followed the instructions or reads sentence aloud.
1 R unable to read sentence aloud due to cognitive impairment.

21.024 [Provide pencil and paper]

Please write a sentence on this piece of paper.

Can you read it to me please?

0 No attempt, only marks on paper, or readable attempt, but not 1.
1 Sentence containing subject and verb.

21.025 *Has R ever been able to write properly?*

0 Is/has previously been able to write.
1 Has never been able to write properly due to lack of education.
2 Probably never been able to write.

21.026 [Show pentagrams. Give pencil and paper.]

Please copy this design.

1 point if pentagrams have 5 sides each and the overlap makes a diamond (four sided figure).

21.027 Sum scores in marked boxes, excluding ratings of 7,8,9 [Max. 30]

Note: Maximum MMS score derived from **21.016** and **21.015** is 5. Only include the better of the two scores from these two items in calculating MMS total score by hand.

If cognitive impairment prevents attempt, enter 00.

21.028 Rate possible underestimation of cognitive abilities on MMS. ☐

0 Probably a fair estimate.
1 Probably an underestimate.
2 Probably marked underestimation.

A MMS Score <=23 or a Verbal Trails rating >= 2 (>45 sec) should be regarded as clear evidence of cognitive impairment or decline (taking into account educational level). A MMS score of 35 coupled with a Verbal Trails rating of 0 (<35 sec) is reasonably good evidence that cognitive impairment is absent. If R shows any evidence of cognitive impairment or decline, regardless of subthreshold MMS or Verbal Trails scores, proceed to **21.030**. Refer to Glossary for further advise on thresholds and suggested MMS parameters that take account of age and education when screening for cognitive impairment.

SKIP => to 21.065 if R is unable to complete MMS or Verbal trails Test due to cognitive impairment.

SKIP => to 22.001 if no evidence of cognitive impairment.

MEMORY

21.029 *Could you again tell me the three objects that I named about 5 minutes ago?* ☐

Score 5 minute recall. Maximum 3.

21.030 ***Subjectively described memory problems*** ☐

Have you had any difficulty with your memory recently?

- *Is it more difficult to remember things than it used to be?*
- *Do you forget where you have left things more than before?*
- *Has the problem become worse recently?*
- *When were you last able to remember things properly*

Do not include trait inattention or 'absent-mindedness' rated 7. Definite loss of memory function, though onset could be earlier.

0 None.
1 Mild difficulty such as 'forgetfulness' that might be due to inattention, worry, etc. [See also **21.126.**]
2 Serious memory loss but R is aware of it. Unlikely to be due sole to inattention, worry, etc.
3 Impairment denied but evident on testing.
4 R unable to answer because of cognitive impairment.

21.031 ***Adoption of methods to compensate for memory deficits***

Do you have to keep notes to remind you of things that need doing, more often than you used to?

Definite change, though onset might have been earlier.

0 None.
1 Mild difficulty that might be due to inattention, worry, etc. [See also **21.126**].
2 Serious memory loss but R is aware of it. Unlikely to be due solely to inattention, worry, depression, etc.
3 R unable to answer because of cognitive impairment.

21.032 ***Forgetting names of intimates***

Do you sometimes forget the name of someone you know very well, like a close friend or a member of your family?

- *When did that begin?*

0 None.
1 Mild difficulty that might be due to inattention, worry, etc.
2 Serious memory loss but R is aware of it. Unlikely to be due solely to inattention, worry, depression, etc.
3 R unable to answer because of cognitive impairment.

21.033 ***Disorientation near home***

Have you been out much (from home or present residence) in past month?

If not:

Why not? Have you been near where you lived but been unable to find your way back home?

- *How did you manage to get back in the end?*

1 One or a few occasions but managed without help.
2 Frequent lapses or required help to get back.
3 R unable to answer, or does not go out, or carers always accompany because of cognitive impairment.

21.034 ***More serious forgetting***

Do you sometimes forget that you have left something cooking or an electric fire on?

- *What happened last time?*
- *Is this happening more often?*

Do not include minor lapses, corrected fairly quickly.

1 One or a few lapses have occurred recently but without serious consequences.
2 Frequent serious lapses or serious consequences.
3 R unable to answer because of cognitive impairment or carers prevent such accidents by supervision.

21.035 ***Immediate recall - repeat numbers***

I am going to say some numbers. Would you repeat them for me please? 6293714

Write the answer down:.........................

Enter the number of digits correct, irrespective of order.

[If movement/willed action disorder rate 8, for this item only.]

For items **21.048 - 21.051, 21.063 - 21.064**, allow plenty of time on tests involving motor skills for those with slow and impaired movements and/or impairment of willed action.

Rate 6 if motor impairment is responsible for inability to complete test correctly.

Rate items **21.036 - 21.062** as follows unless otherwise specified.

0 Test not completed correctly.
1 Test completed correctly.
6 Disorder of movement/willed action responsible for inability to complete task.
8 Task not completed because of some other physical problems.
9 Unable to attempt test because of cognitive impairment.

21.036 *What is the present Prime Minister's [President's] name?*

REMOTE MEMORY

21.037 *What is the date of your birth?*

D ☐☐ M ☐☐ Y ☐☐

Locally and culturally appropriate items of increasing difficulty for direct testing of remote memory may be administered at this point.

LANGUAGE

21.038 *Can you read this for me please?*

[Hand card with sentence: 'People in glass houses shouldn't throw stones'].

0 Cannot read sentence.
1 Some words correct but not whole sentence.
2 Reads the sentence correctly.

21.039 Has R been able to read such a sentence?

0 Is/has previously been able to read.
1 Has never been able read due to lack of education but not 2.
2 Probably never capable of reading because of brain injury, mental retardation, developmental disorder, etc.

21.040 *Tell me the names of as many different animals as you can think of, starting ... now:*

Rate category verbal fluency by recording number correct in 60 seconds:

00, nn, as above.

21.041 [Show three pictures] *What are these objects?*

0 - 3 Record number correct.

21.042 **[Provide pen and paper]** ***Please write a sentence on this piece of paper.***

- *Can you read it to me please?*

0 No attempt, or only marks on the paper.
1 Readable attempt but not 2.
2 Sentence containing subject and verb.

21.043 Has R ever been able to write properly?

0 Probably never capable of writing.
1 Has never been able to write properly due to lack of education.
2 Has previously been able to write.

CALCULATION

21.044 [Show a 10p and a 5p piece (Use local coins)] *How much does this make?* ☐

0, 1

21.045 Has R ever been able to do simple sums? ☐

0 Probably never capable of simple sums.
1 Has never been able to do simple sums due to lack of education.
2 Has previously been able to do simple sums.

21.046 *If someone gives you 15p [3s] as change for £1, how much did you spend?* ☐

0, 1

21.047 Has R ever been able to do simple subtraction? ☐

0 Is/has been previously able to do simple substraction.
1 Has never been able to do simple substraction due to lack of education.
2 Probably never capable of reading because of brain injury, mental retardation, developmental disorder, etc.

PRAXIS

21.048 *Please nod your head.* ☐

0, 1

21.049 *Show me how you wave 'good-bye'.* ☐

0, 1

21.050 *Touch your right ear with your left hand.* ☐

0 Neither correct.
1 Ear or hand correct but not both.
2 Both correct.

21.051 *Can you please try to make your hands look like mine?*

[Touch your thumb to the 5th or most lateral digit of one hand and make an interlocking finger ring by touching the thumb to the most lateral digit of the other].

Rate hand praxis (interlocking rings).

0, 1

ABSTRACTION

21.052 *In what way are an apple and a banana alike?*

0, 1, 2

21.053 *In what way are a boat and a car similar?*

0, 1, 2

PROVERBS: USE CULTURE-SPECIFIC LOCAL EQUIVALENTS WHERE APPROPRIATE

21.054 [Read card if R cannot] *Could you interpret the proverb for me?*

'People in glass houses shouldn't throw stones'.

0 Answer appropriate e.g. don't accuse others if you have faults yourself.
1 Wrong abstract interpretation.
2 Concrete: e.g. 'The glass will get broken'.

21.055 *'A rolling stone gathers no moss'*

Record answer: ....

0, 1, 2 as **21.054**.

21.056 *'Many hands make light work'*

Record answer: ....

0, 1, 2 as **21.054**.

21.057 ***Oddity of Proverb interpretation***

(Rate oddity, not whether interpretation is correct.)

0 No evidence of bizarreness.
1 Answers mostly clear but some degree of bizarreness.
2 Much bizarre content but some degree of clarity.
3 All answers strikingly bizarre.

FUND OF KNOWLEDGE

Ratings for fund of knowledge items.

0 Correct answer within broad limits.
1 Incorrect answer but not 2.
2 Wildly wrong.

21.058 *How tall is the average man approximately?*

0, 1, 2

21.059 *What is the colour of a clear sky?*

0, 1, 2

21.060 *What is the capital of Italy?*

0, 1, 2

21.061 *What is the function of lungs?*

0, 1, 2

21.062 *What is the function of a barometer [or thermometer]?*

0, 1, 2

FRONTAL/SUBCORTICAL TESTS

21.063 *Watch my hand and please try to do what I do.* [Make a fist, then open your hand widely, and then make a scissors-cutting motion using the index and middle fingers.]

Watch my hand again and please try to do what I do this time.

[Perform the following 5 sequences:

Scissors, open hand, fist.
open hand, scissors, fist.
fist, scissors, fist.
scissors, fist, scissors.
fist, open hand, fist.]

Score point for each correct sequence. Maximum 5.

21.064 [Tap the forefinger to your thumb.]

Please tap the forefinger of your thumb like I`m doing.
Whenever I tap once (tap once), I would like you to tap twice (tap twice). And when ever I tap twice (tap twice), I would like you to not to tap at all (spread thumb and index finger).

Ten trials, each with one tap or two taps, in the following sequence:- 1 1 2 2 1 2 1 1 1 2.

Score one point for each correct Finger Tapping trial.
Maximum 10.

Score perseveration at examination at 21.080.

LEVEL OF CONSCIOUSNESS

These items may be rated on observation over a period of time. It is particularly important that the following ratings are made when the subject was unable or unwilling to perform tests.

21.065 ***Level of consciousness***

0 Consciousness clear.

SKIP => to 21.077

1 Abnormality moderate, or severe but transient.
2 Marked and persistent.

21.066 ***Drowsiness***

0 Abnormality not present.
1 Present in moderate or severe degree but transient.
2 Marked and persistent.

21.067 ***Clouding or daze***

Inadequate apprehension of external stimuli and events; impaired attention and orientation; restricted response to questions.

(0, 1, 2 as **21.066.**)

21.068 ***Akinetic mutism***

No voluntary movement, attention held for seconds only, eyes may follow nearby movement, only rare or no speech.

2.056 = Dissociative; **22.006** = Stupor; **21.063 - 21.064** = Abulia.

(0, 1, 2 as **21.066.**)

21.069 ***Coma***

If due to intoxication with alcohol/drugs rate also at **11.031/12.043**.

(0, 1, 2 as **21.066**)

21.070 ***Illusions or visual hallucinations in state of clouded consciousness***

(0, 1, 2 as **21.066**)

21.071 ***Fantastic ideas and delusions in state of clouded consciousness***

Impaired abstraction, comprehension, incoherence, transitory delusions.

(0, 1, 2, as **21.066**)

21.072 ***Psychomotor disturbance in state of clouding***

Excitement, over or under-activity or rapid change, increased reaction time and startle reaction, in state of clouded consciousness.

0 None.
1 Incipient, mild or transient signs.
2 Marked psychomotor symptoms.

21.073 ***Insomnia, reversal, nightmares, more symptoms at night***

Disturbance of the sleep-wake cycle.

(0, 1, 2, as **21.066**)

21.074 ***Acute intoxication on examination***

If due to alcohol or other drug misuse, the attribution should be made at **11.031** or **12.043**. Intoxication from accidental or other poisoning should be rated here. Consider slurred speech, poor coordination, unsteady gait, nystagmus, flushed face.

0 None.
1 Moderate.
2 Marked.

21.075 ***Identify cause of symptoms rated at 21.63***

Exclude substance misuse. Rate at Sections 11 or 12. Enter ICD-10 chapter letter and up to three digits. If none, leave blank.

21.076 ***Fugue***

Wandering away from usual place of residence or work, without clear purpose or apparent intention. Usually associated with daze (**21.067**) and amnesia for period of wandering. Rate dissociative fugue at **2.055**.

OVERALL RATINGS OF COGNITIVE IMPAIRMENT AT EXAMINATION

21.077 ***Memory impairment overall on examination***

0 None.
1 Definite. Moderately severe.
2 Definite and severe.

21.078 ***Loss of intellectual function overall on examination***

0 None.
1 Definite. Moderately sever.
2 Definite and severe.

21.079 ***Disorientation overall (time, place, person)***

0 None.
1 Definite. Moderately severe.
2 Definite and severe.

21.080 ***Perseveration***

0 None.
1 Occasional.
2 Marked.

21.081 ***Retrograde amnesia***

Reduced ability to recall past experiences in reverse order of their occurrence, e.g. the events leading up to concussion.

0 None.
1 Definite. Moderately severe.
2 Definite and severe.

21.082 ***Confabulation***

0 None.
1 Definite. Moderately severe.
2 Definite and severe.

21.083 ***Relatively isolated memory impairment***

Major deficit is in new learning and recent memory, perhaps with confabulation to fill in the gaps. Intellectual functions and more distant memory relatively spared.

0 No memory impairment.
1 Moderate impairment.
2 Severe impairment specifically affecting new learning.

21.084 ***Slowness and difficulty with speech/movement***

Answers brief, long latent period, some questions not answered, little spontaneity, actions slow. **[Always rate section 22.]**

0 Absent.
1 Moderate impairment.
2 Severe impairment.

21.085 ***Rate slowness and difficulty with thought/comprehension***

0 Absent.
1 Moderate impairment.
2 Severe impairment.

21.086 ***Aphasias, apraxias etc.***

0 Absent.
1 Moderate impairment
2 Severe impairment.

21.087 ***Deterioration in self-care on examination***

0 None or not noticeable.
1 Moderate.
2 Very marked.

21.088 ***Uneven distribution of deficits in higher cognitive functions consider results of interviews and evidence from case notes and informants***

0 Absent.
1 Present.

RECENT CLINICAL HISTORY

Information concerning the onset, course and behavioural characteristics of episodes of cognitive decline.

Complete the following items from case records or an informant to supplement the interview and examination with R:

Be overinclusive concerning the definition of cognitive decline. It is necessary to have onset and course information even on marginal cases.

In all the following ratings, 0 = no cognitive decline. If onset in childhood, with lifelong cognitive impairment and no recent decline, or if there is no evidence for cognitive decline, skip to **22.001**.

21.089 ***Length of cognitive decline***

If there is any evidence for cognitive decline, how long has it lasted? Include delirium, amnesic syndrome, cognitive impairment following brain disease or injury, as well as dementia.

1 Less than 3 months.
2 3-6 months.
3 6 months - 1 year.
4 1 - 5 years.
5 Over 5 years.
8 Not known.
9 NA.

21.090 ***Age cognitive decline began***

If there is any evidence for cognitive decline at what age approximately did it begin?

01-94	Record age in years.	97	NK but age 65 or after.
95	Age 95 or over.	98	NK.
96	NK but before age 65.	99	NA.

21.091 ***Pattern of cognitive decline***

If there is any evidence for cognitive decline, rate the pattern of the course?

1 Gradual with more or less continuous development.
2 Rapid development to severe deterioration, including acute delirium, deterioration within 1 month.
3 Intermittent development with periods of exacerbation.
4 Improvement, e.g. after head injury, but some residual impairment.

21.092 ***Transient ischaemic attacks***

Consider focal brain damage, e.g. history of stroke, unilateral spastic weakness of limbs, tendon reflexes, extensor plantar response, pseudobulbar palsy.

1 Gradual, cumulative, series of infarcts.
2 Rapid development to severe impairment within 3 months.
3 Intermittent development with periods of exacerbation.
4 Improvement, e.g. after head injury, but some residual impairment.

21.093 ***Evidence of cerebral damage from investigations***

Consider findings of investigations such as CAT, PET, MRI, SPECT, EEG etc., exclusive of psychometric testing.

1 Cerebral atrophy.
2 Ischemic destruction in the deep white matter of cerebral hemispheres.
3 Evidence of infarcts in cerebral parenchyma.
4 Mixed type (2 and 3).
5 Mixed type (1 and 2 or 3).

21.094 ***History of Hypertension***

0 Absent.
1 Present.

21.095 ***Deterioration in memory***

Rate any deterioration in memory since onset, evident to those who have observed R's activities?

1 Mild, but sufficient to interfere with everyday activities. New material, for example about changes in routine events, much more difficult to learn and recall.
2 Moderately severe. Only very familiar material retained. Serious interference with everyday activities.
3 Severe. Does not remember close relatives. Requires close supervision.

21.096 ***Intellectual deterioration of cognitive abilities***

Rate deterioration in the level of competence of intellectual processes of thinking, reasoning and judgement. Take interference with activities into account.

1 Mild, but sufficient to interfere with activities and be evident to those close to R.
2 Moderately severe. Needs much assistance with shopping, managing money, etc.
3 Severe. Highly dependent. Little intelligent ideation.

21.097 ***Deterioration in self-care***

Ability to look after self, wash, dress, wear appropriate clothes, avoid common dangers, etc.

1 Mild.
2 Moderate.
3 Severe.

21.098 ***Association of dementia with other syndromes***

0 Without additional sympotoms.
1 With predominantly delusional symptoms.
2 With predominantly hallucinatory symptoms.
3 With predominantly depressive symptoms.
4 With mixed symptoms.

21.099 ***Relationship of dementia with delirium***

0 Not with dementia.
1 Superimposed with dementia.

21.100 ***Personality deterioration***

Emotional control, social behaviour, or motivation, amounting to a change in R's personality. Consider marked changes such as emotional lability, shallow euphoria, irritability, disinhibition, suspiciousness, rigidity, neglect of hygiene, altered sexual behaviour.

0 No evidence of deterioration.
1 Mild, but evident to close relatives or friends.
2 Moderately severe. Clearly evident even to those without close contact with R. Some aspects of personality are relatively less affected.
3 Severe damage. Former personality barely recognisable.

SKIP => to 21.103 if 21.100 = 0.

21.101 ***Personality changes occurring within 6 months of viral or bacterial encephalitis***

A reversible syndrome, with symptoms similar to **21.092**, rarely lasting more than 2 years. Residual signs one or more of paralysis, deafness, aphasia, apraxia, acalculia.

0 Absent.
1 Present.

21.102 ***Personality changes occurring within 4 weeks of head trauma with loss of consciousness***

Trauma must occur no more than 4 weeks before onset of at least 3 symptoms - headache, dizziness, excessive fatigue, noise intolerance, emotional change, loss of concentration, insomnia and reduced tolerance for alcohol. See also Section 2 symptoms.

0 Absent.
1 Present.

POSSIBLE CAUSES, PATHOLOGIES OR OTHER INDICATORS ASSOCIATED WITH DEMENTIA

Rate the following:

0 No evidence for the specified cause/pathology.
1 Some evidence for type but uncertain.
2 Reasonable evidence for this type.

21.103 ***Alzheimer.***

21.104 ***Cerebrovascular.***

21.105 ***Pick.***

21.106 ***Jakob-Creutzfeldt.***

21.107 ***Huntington.***

21.108 ***Parkinson.***

21.109 ***HIV.***

21.110 ***Other specified cause of dementia.*** [Enter ICD category at **21.125**].

21.111 ***Unspecified cause of dementia.***

21.112 ***Alcohol causing dementia.***

21.113 ***Substance use causing dementia*** (Type at **21.124**).

21.114 ***Trauma.***

21.115 ***Intracerebral space occupying lesion NOS.***

CAUSES OF DELIRIUM AND AMNESIC SYNDROMES

21.116 ***Amnesic syndrome due to alcohol.***

21.117 ***Amnesic syndrome due to drug*** (Type at 21.124).

21.118 ***Amnesic syndrome NOS.*** (Not alcohol/drugs).

21.119 ***Delirium or toxicity due to alcohol without fits.***

21.120 ***Delirium or toxicity due to alcohol, with fits.***

21.121 ***Delirium or toxicity due to drug, without fits*** (Type at 21.124).

21.122 ***Delirium or toxicity due to drugs, with fits*** (Type at 21.124).

21.123 ***Delirium NOS.***

21.124 ***Drug rated at 21.113, 21.117, 21.121, 21.122 as possible casual agent identify:***

1 = OPD, 2 = CAN, 3 = SED, 4 = COC, 5 = STM, 6 = HAL, 7 = PCP, 8 = VOL, 9 = OTH.

21.125 ***Identify organic cause of cognitive impairment rated in Section 21***

Specify condition, not specified elsewhere, known to have caused cognitive impairment rated in this Section. Optional item specific attribution can be made in Section 20.

Specify ICD-10 chapter and number:

[e.g. head trauma, hydrocephalus, subdural hematoma, hypothyroid, Vitamin B-12 or folic acid deficiency, niacin deficiency, hypercalcemia, neurosyphilis. Enter identifying letter for ICD-10 chapter and up to 3 digits. If none, leave blank.]

21.126 ***Rate interference with activities due to Section 21 impairment***

0 None.
1 Little.
2 Moderate/intermittent.
3 Severe/incapacitating.

OPTIONAL CHECKLIST (Items 21.127 - 21.134)

MILD COGNITIVE DISORDER

Rate each item:

0 Absent.
1 Present.

21.127 ***Subjective forgetfulness for recent events*** (minimum 2 weeks).

21.128 ***Poor concentration and attention*** (minimum 2 weeks).

21.129 ***Subjective difficulty with new learning*** (minimum 2 weeks). ☐

21.130 ***Slowing in problem-solving or abstraction.*** ☐

21.131 ***Difficulty in comprehending and word-finding.*** ☐

21.132 ***Impaired visuospatial functioning.*** ☐

21.133 ***Evidence of infection or cerebral or systemic disorder closely preceding or following symptoms.*** ☐

21.134 ***No objective evidence of neurological dysfunction and laboratory tests all normal.*** ☐

21.135 ***Performance decline or abnormality in quantified cognitive assessments.*** ☐

OTHER MENTAL DISORDERS RATED ELSEWHERE IN THE PSE, DUE TO CEREBRAL DISEASE, DAMAGE OR DYSFUNCTION

The category covers disorders such as Organic hallucinosis, Organic catatonia and Organic anxiety disorder. They are rated with their causes in the appropriate sections of the PSE. The following criteria should be applied as far as possible:

(a) Evidence of cerebral disease, damage or dysfunction, whether direct or via non-cerebral disorders.
(b) A temporal relationship between the development of disease and the onset of the mental disorder, such as onset within 3 months (before or after) the cerebral disorder.
(c) Recovery or improvement after removal of the cause (a difficult criterion to meet).
(d) No other obvious cause.

If alcohol or drugs cause mental disorders, e.g. 'alcoholic or drug-induced hallucinoses', the attribution is entered in Section 11 or 12, and, if required in specific projects at an item specific level in Sections 13 and 20 (optional).

See Sections 13 and 20, and Glossary for full description.

EXAMINATION FOR MOVEMENT DISORDERS

This examination is suitable for screening purposes only. It covers both psychiatric and neurological disorders and supplements the disorders of movement and speech (Sections 22 and 24) rated on the basis of the whole clinical examination. Finer detail will be provided in a Module being prepared by WHO. Some of the following items, such as the first, can be taken at other points in the interview. Side effects of medication can be rated on the **AIMS** scale and the **TAKE** scale.

Observe face, head, neck, trunk and limbs during the following exercises for abnormalities of posture and movement.

Offer to shake hands

If an unusual reaction, consider negativism, ambitendance, forced grasping, autistic spectrum.

Flex and extend elbows gently

If unusual tone, consider flexibilitas, spasticity, freezing.

Ask respondent to stand still

Consider akathisia.

Ask R to extend arms in front of body, palms down

Consider tremor, choreic movement, athetosis, posture, axial hyperkinesis.

Ask R to touch examiner's finger and then own nose

Arm should be completely extended, i.e. examiner should stand sufficiently far away. Consider intention tremor.

Ask R to imitate facial movements and expression

Examiner should smile, frown, wink, blow out both cheeks. Consider ability to imitate facial expressions and rapidity of change from one to another. Consider facial immobility.

Ask R to open mouth and protrude tongue

Consider facial and oral movement. Drooling. State of teeth and dentures.

Ask R to walk across room then return

Consider gait.

RATING SCALE III Behaviour, speech and affect

Consider **21.065 - 21.076,** behavioural items for cognitive impairment.

Most items in sections **21 - 23** are rated on a 3-point scale (0-2) on the basis of severity and frequency during the **past month**. Information from records for the period may be used as well.

The examination should be supplemented by taking into account any other observations of relevance, e.g. in case records or information from professionals or relatives. Severe behavioural abnormalities may not be observed at examination because of the short time sample, but when they **are** present, skilled direct observations are of great importance. The items listed are also worth rating because of their possible juxtaposition with other symptoms. The time period rated is the month before examination.

Many behavioural items are also included in Item Groups and can be rated in the Checklist.

- 0 Behaviour not present during past month.
- 1 Unequivocally present during past month, moderate severity only. Use all information available.
- 2 Present in severe form during past month or at examination.
- 8 Unsure whether present or not after adequate examination.
- 9 Not appropriate to make a rating because examination or records are incomplete, or due to physical factor.

If present but partly or wholly attributable to medication, psychoactive substance or physical condition rate attribution of cause in Section 20.

Section 24 is a brief schedule for rating abnormal social behaviour over the past month using information from examination, case records and informants.

22 Motor and behavioural items

Use Rating Scale III

Behaviour relating to cognitive impairment is rated in Section 21.

REDUCED MOTOR ACTIVITY

22.001 ***Slumped Posture***

Bowed head, round shoulder unusual for age.

22.002 ***Stillness***

Sits or stands abnormally still. Very little movement unless asked to move.

22.003 ***Slowness of movement***

Unusual for age and physical condition. Motor retardation.

22.004 ***Delayed start to movements***

Delay in beginning and performing movements.

22.005 ***Underactivity***

Requires external stimulus in order to start moving and then tends to stop as soon as possible.

22.006 ***Stupor***

Total or nearly total lack of spontaneous movement and marked decrease in reactivity to environment.

1 Fluctuates.
2 Dominates clinical picture.

22.007 ***Slow speech***

Long pauses before answering and long pauses between words or sentences. (Do not include drawling or slow articulation).

22.008 ***Shuffling gait***

Unusual for age.

22.009 ***Arms not swung***

Does not swing arm when walking.

22.010 ***Negative posture***

1 Suggests passive aversion, e.g. eyes closed, hunched, head bowed, etc.
2 Suggests active aversion, e.g. head turned away, avoids eye contact, back turned.

22.011 ***Gesture***

1 Little use of gesture to accompany speech or illustrate meaning.
2 Virtually no use of gesture at all.

22.012 ***Facial expression***

1 Little change in expression to accompany or illustrate meaning or feeling.

2 Virtually expressionless face gives no clue to inner thoughts or feelings.

22.013 ***Expression in Voice***

1 Voice changes little in expression with change of topic.
2 Voice monotonous and expressionless throughout all speech.

22.014 ***Eye-contact***

1 Eye contact used unskilfully, e.g. stares, gives wrong cues, etc.
2 Makes almost no eye contact at all or only inappropriate.

INCREASED MOTOR ACTIVITY

22.015 ***Distractibility***

Changes behaviour or speech inappropriately though attending to irrelevant noises or events or objects.

22.016 ***Restlessness***

1 Noticeably restless, not amounting to 2.
2 Constantly fiddling, changing position, standing or sitting down, etc. (Not as severe as **22.015**).

22.017 ***Agitation***

1 Pacing up and down, moving about, not completing actions, etc., but periods of normal activity.
2 Constant purposeless agitation.

22.018 ***Overactivity or excitement***

Apparently purposeless motor activity, not influenced by external stimuli.

1 Intermittently runs or jumps, waves arms wildly, shouts, screams.
2 Constant excitement dominates clinical picture.

22.019 ***Rapid alternation between stupor and excitement***

0 Not present.
1 Less than 2 weeks.
2 2 weeks or more.

22.020 ***Destructive or violent***

1 Destructive to objects.
2 Violent to people (or violently destructive).

Aggressive attitude rated at **23.004**. See Glossary.

POSSIBLE CATATONIC BEHAVIOUR

Items **22.021 - 22.029** should be rated according to the time the symptoms have been present:

1 Less than 2 weeks.
2 2 weeks or more.

22.021 ***Negativism***

Does opposite to what is asked.

22.022 ***Ambitendence***

Begins to take hand then withdraws, etc. Unable to complete movement e.g. passing through a door (advances, withdraws, advances again, etc.).

1 Less than 2 weeks.
2 2 weeks or more.

22.023 ***Forced grasping***

Takes hand repeatedly. Does so when asked not to.

1 Less than 2 weeks.
2 2 weeks or more.

22.024 ***Echopraxia***

Echolalia, imitates movements repeats phrases.

1 Less than 2 weeks.
2 2 weeks or more.

22.025 ***Flexibilitas cerea***

Distinguish from resistance. Lead pipe rigidity. Arms remain where placed for 15 seconds or more.

1 Less than 2 weeks.
2 2 weeks or more.

22.026 ***Opposition***

Movement in any direction is countered by equal resistance in the opposite direction.

1 Less than 2 weeks.
2 2 weeks or more.

22.027 ***Jerkiness***

Lack of smoothness of voluntary movements for 15 seconds or more.

1 Less than 2 weeks.
2 2 weeks or more.

22.028 ***Freezing***

Arm remains in one position during voluntary movement.

1 Less than 2 weeks.
2 2 weeks or more.

22.029 ***Automatic obedience***

Fingertip pressure on any part of the body produces movement in that direction.

1 Less than 2 weeks.
2 2 weeks or more.

22.030 ***Catatonic symptoms dominate clinical picture***

0 None.
1 Present but not dominant.
2 Dominates the clinical picture.

22.031 ***Dates of PERIOD of Section 22 symptoms*** to

Recording the dates of abnormalities in Section 22 will depend on reliable information from informants. It is necessary to record only if dates of ratings in this Section are different from PERIOD rated in Section **1** (items **1.007, 1.009, 1.010 - 1.013**).

22.032 ***Interference with activities due to Section 22 symptoms***

Behavioural problems can cause considerable interference with activities. If the respondent is living in an institution or other sheltered setting, there may be fewer opportunities to assess the effect such behaviours have on functioning, and yet they may be the major reason for loss of independent living in the community; if so rate 2.

You have mentioned [summarise symptoms] *during PERIOD. Overall, how much interference has there been with your everyday activities because of these problems?*

Rate interference due to symptoms in Section 22.

0 No Section 22 symptoms present in significant degree.
1 Symptom/s present but little interference.
2 Moderate or intermittent interference.
3 Severe to incapacitating interference.

Rate attribution of physical (including drug, substance etc) and or psychosocial cause at Section 13/20.

Rate stupor at **22.006**; excitement at **22.018**.

Rate attribution of cause of catatonic symptoms.

Although opportunity to rate attribution of physical (including drug, medical condition, etc) cause is present using the optional attributional scale and the dashed boxes, items **22.033**and **22.034** allow the user to re-rate organic cause at the section / syndrome level.

Consider disorders in the autistic spectrum. Asperger's syndrome, is particularly associated. Mid-brain lesions, and post-encephalitic Parkinsonism should be considerd. Make all necessary ratings in Section 20.

22.033 ***Organic cause of catatonic symptoms***

0 Absent.
1 probable organic cause; not fully confirmed.
2 Definite; confirmed by expert investigation.
8 Uncertain whether organic or not.

22.034 ***Identify the organic of cause of catatonic symptoms***

Consider disorders in the autistic spectrum. Asperger's syndrome, is particularly associated. Mid-brain lesions, and post-encephalitic Parkinsonism should be considered. Enter the letter identifying the ICD10 chapter and 1 - 3 digits. If none, leave blank. Make all necessary ratings in Section 20.

ODD BEHAVIOUR OR APPEARANCE

22.035 ***Odd or inappropriate appearance***

Odd clothes, ornaments etc. Would stand out in a social group because of markedly inappropriate appearance.

22.036 ***Embarrassing behaviour***

Making sexual advances, loss of social restraint e.g. scratches genitals, passes loud flatus, etc.

22.037 ***Bizarre behaviour***

Secret documents or codes. Special ornaments or clothes, with idiosyncratic meanings. Corks in ears, etc. Make-up bizarrely applied. Appearance grossly eccentric and embarrassing.

22.038 ***Irreverent behaviour***

Sings. Makes facetious jokes out of context. Flippant. Unduly familiar.

22.039 ***Histrionic behaviour***

Feelings expressed in exaggerated, dramatic manner.

22.040 ***Self neglect***

Clothes inadequate for warmth and protection, irrespective of whether odd or embarrassing. Unkempt. Dirty. Smelly. Unshaven.

22.041 ***Neglect of health care***

Inadequate diet. Ignores need for shelter. Grossly neglects physical disabilities or diseases or attention to dental hygiene: caries, ill fitting dentures, does not use dentures.

22.042 ***Personal hygiene and habits*** ☐

Spits, smears, eats rubbish.

22.043 ***Self injury*** ☐

Head banging, picking sores, biting self, etc.

22.044 ***Creates chaos*** ☐

Moves and scatters objects aimlessly. Disjointed behaviour.

22.045 ***Neglect of common dangers*** ☐

Does not understand common dangers, e.g. wanders across busy roads, not to be trusted with gas taps, electric fires, etc.

22.046 ***Hoarding objects*** ☐

Collection of old newspapers or useless objects.

22.047 ***Complex mannerisms*** ☐

Odd stylised and idiosyncratic movements or actions, e.g. tapping foot four times before entering a doorway.

22.048 ***Posturing*** ☐

Odd postures of parts of body, e.g. 'Balinese dancer' posture of hands and fingers.

22.049 ***Simple stereotypies*** ☐

Repetitive movements. Rocking, rubbing, nodding, swaying, twisting objects, tapping, feeling surfaces.

22.050 ***Inane giggling or inappropriate laughter*** ☐

Must be non-congruent with social or emotional context.

22.051 ***Apparently hallucinating behaviour*** ☐

As though hears voices or sees visions; e.g. lips move, looks round, replies, etc.

Such behaviour does not necessarily imply hallucinations.

22.052 ***Behaviour associated with single interest***

The interest must exclude or modify almost every other form of activity so that it appears markedly odd.

MOTOR AND BEHAVIOURAL ITEMS 287

23 Affect

Use Rating Scale III throughout.

23.001 ***Observed anxiety***

Fearful apprehensive look, frightened tone of voice, tremor in voice/hands, autonomic signs.

23.002 ***Observed depression***

Sad, mournful look, tears, gloomy tone of voice, deep sighing, voice chokes on distressing topic.

23.003 ***Elated affect***

Abnormally cheerful, smiling, euphoric, elated.

1 4 days minimum.
2 1 week minimum.
3 2 week minimum.
4 4 or more weeks.

23.004 ***Hostile irritability***

Unco-operative, irritable, angry, overtly hostile, discontented, haughty, antagonistic. Aggressive, but not violent, rate 2.

23.005 ***Perplexity***

23.006 ***Lability of mood***

Either with one affect or with emotional turmoil, i.e. rapid shifts from one mood to another without the persistence of any one affect. Rate (1) or (2) according to the frequency of change and briefness of moods (see glossary).

23.007 ***Emotional outbursts***

Uncontrolled, intense and frequent emotional outbursts.

23.008 ***Blunting or flattening of affect***

Decrease in emotional responsiveness shown in facial expression, tone of voice, use of gesture, etc.

1 Blunting. Quantitative decrease compared with expectation.
2 Severe and uniform flatness of affect.

23.009 ***Incongruity of affect***

Emotion is shown, but not congruent with topic. Parathymia.

23.010 ***Suspiciousness*** [Check **3.011**]

Rate on observations during whole interview or period. Dates and duration of PERIOD/S of abnormal affect should be rated in Sections 6 and 10 as interviewer is unlikely to have *observed* onset directly.

24 Speech abnormalities

Use Rating Scale III throughout.

24.001 ***Abnormal loudness of voice***

Inappropriate to context.

24.002 ***Abnormal quietness of voice***

Inappropriate to context.

24.003 ***Mumbling 'to self'***

Talks or mutters, or whispers out loud. Words not distinguishable.

24.004 ***Pressure of speech***

More copious than usual. May be too rapid and too loud. See Glossary for full definition.

24.005 ***Flight of ideas***

Words are associated together inappropriately by sound or rhyme or clang association.

A rating of 2 means that very little normal speech is present.

24.006 ***Circumstantiality***

Gets lost in insignificant detail. Cannot differentiate what is essential and therefore loses the thread of the answer. Gives far more detail about some particular topic (or fragment of a topic) than is necessary.

24.007 ***Rambling speech***

'Drivelling', 'Faseln'. Rambling on in a vague, muddled way, beginning more or less on the point but gradually wandering far from it.

24.008 ***Perseveration***

Words or phrases repeated so as to be meaningless. Extreme form is verbigeration, rate 2.

24.009 ***Rumination***

Endless preoccupation with one theme. Mulling over a topic to no purpose.

24.010 ***Approximate answers***

R understands the question and answers in a related but incorrect way, e.g. gives a completely wrong date or place.

24.011 ***Neologisms and idiosyncratic use of words or phrases***

For examples see Glossary definitions.

24.012 ***Incoherence of speech***

The subject's meaning is obscured by distorted grammar, lack of logical connection between one part of a sentence and another or between sentences.

A rating of 2 means that very little normal speech is present.

24.013 ***Magical or markedly illogical thinking***

24.014 ***Blocking***

R stops talking and then begins again on same or different topic. Not distraction, lapse of attention, or lack of understanding.

24.015 ***Poverty of content of speech***

The subject talks freely but so vaguely that little information is given in spite of the number of words used. Exclude incoherence or flight of ideas. Rate only if severe and always give written example. Slowness of speech is rated at item **24.007**. See Glossary for examples.

24.016 ***Restricted quantity of speech***

Subject frequently fails to answer, questions have to be repeated, restricted to minimum necessary, no extra sentences, no additional comments.

24.017 ***Muteness***

1 Very few words; fewer than 20 in all.
[Unless Yes or No, all appropriate]
2 Almost or totally mute.

24.018 ***Nonverbal communication during speech***

Lack or under-use of normal non-verbal gestures, facial expressions, changes of tone, pitch and loudness, eye-contact, etc., during conversation.

24.019 ***Ability to keep conversation going***

No small talk. Does not initiate or keep a conversation going.

24.020 ***Rewards in conversation***

Does not reward other person for speaking by non-verbal acknowledgements, nodding, smiling, *'I see'*, etc. (See also **24.010 - 24.014**).

24.021 ***Misleading answers***

Subject's answers are misleading because answers *'yes'* (or *'no'*) to everything, or frequent self-contradictions, or appears to be deliberately misleading. Do not include incoherence, flight of ideas or poverty of speech here.

24.022 ***Non-social speech***

Non-social speech. Talks, mutters, whispers out loud, out of context of conversation.

24.023 ***Grossly pedantic speech***

Over-formal, over-technical terms or phrases that sound very odd but are used without understanding that they are inappropriate. See Glossary definition.

24.024 ***Delayed echolalia and repetitive speech***

Uses phrases obviously said to R in second person. E.g. *'Do you want a biscuit'* means *'I want a biscuit'*.

24.025 ***Ungrammatical speech in real attempt to communicate***

E.g. *'I did went on bus'*. Not due to lack of education. May be associated with pronominal reversal: *'You did went on bus'*.

24.026 ***Made-up words in real attempt to communicate***

When the correct word is unknown e.g. *'tram-bus'* for *'trolley- bus'*; *'earring plugs'* for *'headphones'*.

24.027 ***No 'theory of mind'***

Lack of awareness that other people have thoughts, ideas and feelings. See Glossary.

24.028 ***Evasiveness***

Avoids answering direct questions, particularly those concerned with abnormal experiences. May or may not give misleading answers, which is a different characteristic, see **24.023.**

24.029 ***Lack of direction or purpose***

Consider complete interview and past month. How far does R show evidence of initiative, passive or active interest in any kind of occupation, plans, interest in topics in the interview, etc.

0 Normal purposefulness.
1 Some active interests but these stand out against a general background of passivity.
2 Interest mainly passive. Very little active initiative or purposefulness shown.
3 Completely apathetic. Needs to be stimulated by others to engage in activity and even then does not persist.

24.030 ***Insight into positive Part II symptoms***

R has been asked, at several points, about the explanation for various groups of symptoms. Rate here overall insight into the nature of psychotic symptoms more generally, including associated behaviour.

0 No psychotic symptoms.
1 Good insight into the mental or other natural origin of symptoms.
2 Some insight but fluctuating or guarded. Include evasion or vagueness here.
3 Very little insight. Explanations mainly couched in delusional terms.

24.031 ***Adequacy of ratings in sections 22 - 24***

Use Glossary checklist of difficulties likely to be met.

0 No difficulties in making ratings.
1 Some problems but most ratings reasonably accurate.
2 Major difficulty in conducting examination. Ratings must be interpreted with caution.
3 Impossible to rate PSE10 with any confidence.

Adequacy of Part I is rated at **13.129.**
Adequacy of part II is rated at **20.092.**

25 Social impairment

This Section should be used if the interview as a whole, and items **24.025 - 24.029** in particular, suggest the presence of social impairment. Use information gathered at interview, from case records and from informant. The period covered is the past month. Leave blank if not completed.

Rate each item:

0 Does not apply.
1 Applies somewhat.
2 Applies markedly.

25.001 ***Markedly stiff, over-formal, stilted manner.*** ☐

25.002 ***Regarded as very eccentric by peers.*** ☐

25.003 ***Odd attachment to one or a few particular objects.*** ☐

25.004 ***Wishes to be sociable but fails but to make relationships with peers.*** ☐

25.005 ***Can only cope with company on his/her own terms.*** ☐

25.006 ***Has no real friend.*** ☐

25.007 ***Lacks empathy.*** ☐

25.008 ***Has little or no idea of how to cooperate in a team, or to compete against another team.*** ☐

25.009 ***Conspicuously lacks common sense.*** ☐

25.010 ***Has clumsy, poorly coordinated, odd and ungainly movements.*** ☐

25.011 ***Tends to be exploited or bullied.*** ☐

25.012 ***Seems to live in a world of his/her own, intensely preoccupied with a few idiosyncratic interests. But can't cope with everyday matters.*** ☐

25.013 ***Surprisingly good at a few isolated subjects.*** ☐

25.014 ***Accumulates facts on specific subjects (good rote memory) but does not really understand the meaning or make use of the knowledge.*** ☐

25.015 ***Has an odd style of communication with a formal, fussy, pedantic or 'robot-like' language.*** ☐

25.016 ***Uses language freely but fails to adjust phrases to fit social context or take account of the interests of different listeners.*** ☐

25.017 ***Makes naive and embarrassing remark.*** ☐

25.018 ***Marked impairment of non-verbal means of expression, monotonous tone of voice, little facial expression, no (or inappropriate) gesture, poor use of eye contact.*** ☐

SKIP => to Item Group Checklist or Clinical History Schedule.

26 The Item Group Checklist

INTRODUCTION

In the international field trials of SCAN, it was found that respondents interviewed during active episodes of mental disorder were usually able to provide detailed information not only about 'present state' but also about the previous clinical history. However, there are obvious exceptions; see page 6 (in PSE-10) for possible contingencies.

The Item Group Checklist (IGC) provides a simple means of rating information obtained only from case-records and/or informants other than the respondent. The IGs are not diagnostic syndromes; they cannot be translated directly into ICD10 disorders but must be processed by a version of the CATEGO-5 program adapted from the one that is applied to data from PSE10. The resulting classification is approximate compared with that from the PSE but it is a useful supplement to PSE information, and can substitute in situations where the PSE cannot be fully completed.

The 40 Item Groups included in the Checklist cover disorders in subchapters F2 (psychotic disorders), F3 (affective disorders) and F40 - 42 (neurotic disorders) of ICD10. Disorders in subchapters F0 (cognitive impairment), F1 (substance use disorders), F43 - 45 (stress, dissociative and somatoform disorders) and F5 (appetite and sleep disorders and sexual dysfunctions), which are difficult to rate without interviewing or examining the respondent, are not included. It may occasionally be appropriate (particularly for the eating and substance use disorders) to rate the PSE itself from case-records or an informant.

Users of the IGC must have been trained in the use of the PSE and its Glossary, and be completely familiar with the structure of SCAN. This provides a substantial degree of operationalisation for rating IGs, each of which is composed of designated PSE-10 items.

INDICATIONS FOR USE

There are three main indications for using the IGC:

1 When the Respondent can provide information about present state but is unable or unwilling to remember or describe accurately the events or symptoms of a previous episode.

In such a case, the previous period (RE or LB) is chosen and dated in the same way as for the PSE.

However, it is not necessary, for most purposes, to use IGC to rate a past episode if the symptoms then were very similar to those in the present state.

2 When an interview with the respondent is impossible. In this case, the examiner chooses and dates PS, or PS + RE or PS + LB. The period/s chosen will depend on a clinical judgement as to which will most adequately cover the history for purposes of diagnosis. Both periods can be rated with IGC if no other method is possible. Often there will be an opportunity to observe behaviour, speech and affect on examination, even when an interview is impossible. These observations can be used, with other information, to rate **26.048 - 26.057** (IG 31 - IG 40).

3 For use in research projects that require the maximum use of information in case records. Even when two PSEs have been completed, extra IGCs can be processed separately as a supplement to the routine package. Also, up to 6 episodes can be rated using PSEs, IGCs, or a mixture of the two.

Duplicate SCAN face sheet

Do not complete if PSE-10 has been used.

This Face Sheet is identical to the one on page 7. It is included for convenience when the IGC is used without any PSE.

00.001 ***Version of SCAN.*** [ ]

00.002 ***Project/Centre number.*** [ ][ ]

00.003 ***Respondent number.*** [ ][ ][ ][ ]

00.004 ***Rater number.*** [ ][ ][ ]

00.005 ***Date of first PSE examination.*** [ ][ ][ ][ ]

[If no examination, date of completion of first schedule]

SKIP => to 0.016 if non-routine option is chosen.

Only 2 Periods may be rated in the routine option: 2 x PSEs, 2 x IGCs, or 1 PSE + 1 IGC.

Period(s) rated using PSE(s) - any source of information.

0 PSE not rated.
1 PSE rated for the period stated.

00.006 ***Present state (PS).***

00.007 ***Representative previous episode (RE).***

00.008 ***Lifetime before PS (LB).***

00.009 ***This PSE completed by:***

0 PSE not used (IGC used instead).
1 Interviewer rating own interview.
2 Observer rating interview.
3 Rating from videotape.
4 Rating from audiotape.
5 Other, Specify ....

00.010 ***Source of information used to rate PSE.***

[Enter for PS in left box, for RE or LB in right box].

0 PSE not used [Only IGCs].
1 Respondent only.
2 Respondent and other sources.
3 Case records only.
4 Informant/s only.
5 Case records and informant/s.
Specify relative, professional, etc. ....
6 Prepared abstract, ...

Period rated using IGC(s)

0 IGC not used.
1 IGC used for the period stated.

00.011 ***Present state (PS).***

00.012 ***Representative previous episode (RE).***

00.013 ***Lifetime before (LB).***

)0.014 ***Source of information used to rate IGC.***

[Enter for PS in left box, for RE or LB in right box].

0 IGC not used.
1 Respondent only [e.g. observed behaviour ratings only].
2 Respondent and other sources.
3 Case records only.
4 Informant/s only.
5 Case records and informant/s.
Specify relative, professional, etc. ....
6 Prepared abstract, ...

Clinical History Schedule

[This is recommended if a narrative output is required.]

00.015 ***Use of Clinical History Schedule.***

0 Not used.
1 Used.

Use of the non-routine option (Episode list; see items 1.019 -1.024).

00.016 ***Identifying number of period rated.***

0 Not used.
1-6 Number of episode rated in this IGC schedule.

Indicate the form of SCAN data recording used:

[Rate one option only.]

0 Not used.
1 Used.

00.017 ***Rating boxes in SCAN schedules [PSE, IGC, CHS]***

00.018 ***Coding booklet (fixed format).***

00.019 ***Recording booklet (free entry).***

00.020 ***Computer.***

SOCIODEMOGRAPHIC ITEMS

If there is any possibility of cognitive impairment, the following items may be completed as a test, by asking R for the information. Skip to Section **21** if necessary.

[CHS has a more comprehensive social history, which can be completed with the respondent]

00.021 **Age** [If not known, leave blank]

00.022 **Date of birth** [Day, Month and Year]

00.023 **Sex:**

1 Male.
2 Female.

00.024 ***Marital status:***

1 Married.
2 Cohabiting.
3 Separated.
4 Divorced.
5 Widowed.
6 Never married and not cohabiting.

00.025 ***Number of children born alive.***

00.026 ***Number of years of formal schooling.***

00.027 ***Occupational status:***

1 Full time paid.
2 Part time paid.
3 Household duties.
4 Unemployed.
5 Retired.
6 Sheltered work.
7 student.
8 NK.
9 NA (e.g. Maternity leave).

00.028 ***Nature of paid employment, specify:*** **...................**

00.029 ***If in paid employment: Number of months working during the past year.***

DATES OF PERIODS CHOSEN

Enter the dates of the chosen period for rating IGC, using the same principles as for the PSE. The principles are described in items **1.007 - 1.013** on page 13.

If PSE is the chosen period, enter:

26.001 *Length in months of present episode* ☐☐☐

[See item **1.007** for definition.]

26.002 *Number of days in PS* ☐☐

[This period should usually be 28 days. See item **1.008.**]

If RE is the chosen period, enter:

26.003 *Dates of representative previous period* ☐☐☐☐ to ☐☐☐☐

[A discrete single period which, together with P, adequately represents the clinically significant symptoms during the course. See item **1.009.**]

If LB is the chosen period, enter dates of constituent LB episodes:

[Date LB from the first onset of disorder to the start of PS. See instructions for items **1.010 - 1.013.**]

26.004 *Neurotic.* ☐☐☐☐ to ☐☐☐☐

26.005 *Depressive.* ☐☐☐☐ to ☐☐☐☐

26.006 *Manic.* ☐☐☐☐ to ☐☐☐☐

26.007 *Psychotic.* ☐☐☐☐ to ☐☐☐☐

If IGC Is used to rate LB:

Enter approximate age of first onset of any episodes of symptom types dated in **26.004 - 26.007**. [See items **1.014 - 1.017.**]

26.008 *Neurotic.* ☐☐

26.009 ***Depressive.***

26.010 ***Manic.***

26.011 ***Psychotic.***

SPECIAL EPISODE LIST

See items **1.019 - 1.024.** Any mixture of PSE and IGC schedules may be used but each will be processed as an individual case. If any PSE is used, enter initial details on the first Face Sheet, page 11.

26.012 ***Episode 1*** to

26.013 ***Episode 2*** to

26.014 ***Episode 3*** to

26.015 ***Episode 4*** to

26.016 ***Episode 5*** to

26.017 ***Episode 6*** to

THE STRUCTURE OF IGS AND RATING PROCEDURE

Ratings of item groups should be entered in fixed format on the coding sheets, in the Record Book, or directly into the boxes provided in the SCAN schedule (including the computer version).

The items constituting each Item Group are listed in the Checklist with their names. They provide a general guide to the content of each IG. The PSE item definitions, and criteria for presence and severity, should be used as models when rating the presence of IGs. The threshold for rating a symptom present should be at least as high as that set in the PSE for the relevant item. These levels are specified against each item and raters are expected to be familiar with their clinical meaning as defined in PSE10. If in doubt, rate 8.

It would occasionally be possible, if informants or case records were able to provide sufficient information, to score the IGs by rating each constituent symptom. This practice is not recommended, since in such a case the ratings would more easily be entered using the PSE.

Each Item Group is given an overall rating based on the number of constituent symptoms present during the period and the severity of the distress or disablement that accompanies them. Clinical judgement is required in all cases, because the information will often be scanty. In addition, the criteria for rating specified Rating Scale IV must sometimes be relaxed on clinical grounds. For example, IG7 (**26.024**) is composed of five types of obsessional symptoms, but any one of them can be totally disabling (therefore rate 2). And if one type is present, and associated with only moderate disablement, it is still proper to give a rating of 1. The instructions for rating must be interpreted with clinical common sense.

A count of the number of IGs rated 8 (unsure whether present or absent) forms part of the output. Free use should be made of this rating, for example if the presence of psychotic or 'physical' symptoms make the rating of non-psychotic IGs difficult. Traits are not considered when rating IGs. Since the information will sometimes be scanty, it is important to follow the general SCAN rule: 'When in doubt, choose the less definite or less severe rating'; 0 rather than 8, 8 rather than 1, 1 rather than 2.

RATING SCALE IV

0 Absent. Below IG threshold.
1 Subject to clinical discretion, at least two items should definitely pass PSE threshold level, PLUS at least moderate disability (interference with everyday activities) and/or moderate or severe distress resulting from the symptoms.
2 As (1), but with severe degree of distress and/or disability. Usually at least three items pass PSE threshold but this is at the interviewer's clinical discretion.
8 Uncertain whether IG is present (usually because of insufficient information).
9 Information incomplete or records inadequate.

LIST OF ITEM GROUPS WITH CONSTITUENT ITEMS

26.018 *Nervous tension (IG1)* ☐

2.009	Satisfaction with care	2 - 3
3.001	Worrying	2 - 3
3.002	Nervous tension	2 - 3
3.008	Sensitivity to noise	2 - 3
3.009	Irritability	2 - 3
4.024	Anxious foreboding	2 - 3
6.015	Loss of self-confidence	2 - 3
6.016	Social withdrawal	2 - 3
8.010	Takes sleeping tablets	2

8.011	Delay going to sleep	2
8.017	Nightmares	2

26.019 ***Muscular tension (IG2)*** ☐

3.003	Muscular tension	2 - 3
3.005	Localised tension pains	2 - 3
3.006	Restlessness	2 - 3
3.007	Fatiguability and exhaustion	2 - 3

26.020 ***Autonomic anxiety and panic (IG3)*** ☐

3+ positive items from **4.003 - 4.019** are required, as well as 1 - 2 other items.

4.003	Breathlessness	1
4.004	Tachycardia	1
4.005	Dizziness or faintness	1
4.006	Tingling or numbness	1
4.007	Tightness in chest	1
4.008	Mouth feels dry	1
4.009	Difficulty swallowing	1
4.010	Excessive sweating	1
4.011	Tremor of hands or limbs	1
4.012	Hot or cold flushes	1
4.013	Feeling of unreality	1
4.014	Churning stomach	1
4.015	Fear of death or dying	1
4.016	Feeling of choking	1
4.017	Fear of losing control	1
4.019	Other irrational fears	1
4.020	Panic attacks	see item
4.023	Persisting free-floating anxiety	2 - 3
4.026	Depersonalisation with anxiety	2 - 3
4.055	Panics without phobias	3
23.001	Anxiety on examination PS	1 - 2

26.021 ***Agoraphobia (IG4)*** ☐

1+ positive item from **4.027 - 4.032** is required, as well as **4.046**.

4.027	Anxiety in open spaces	1
4.028	Anxiety when no easy exit	1
4.029	Anxiety when travelling alone	1
4.030	Anxiety when away alone	1
4.031	Fear of collapsing while alone	1
4.032	Anxiety when alone indoors	1
4.046	Severity of situational phobias	2 - 3

26.022 ***Social Phobia (IG5)*** ☐

1+ positive item from **4.033 - 4.036** is required, as well as **4.049**.

4.033	Anxiety when appearing in public	1
4.034	Excessive blushing in company	1
4.035	Anxiety in a small social group	1
4.036	Anxiety about speaking to others	1
4.049	Severity of social phobias	2 - 3

26.023 ***Specific phobia (IG6)*** ☐

1+ positive item from **4.037 - 4.044** is required, as well as item **4.052**.

4.037	Fear of severe weather	1
4.038	Fear of water	1
4.039	Fear of small animals	1
4.040	Fear of enclosed spaces	1
4.041	Fear of flying	1
4.042	Fear of heights	1
4.043	Fear of medical situations	1
4.044	Fear of contracting illness, aids, etc.	1
4.052	Severity of specific phobias	2 - 3

26.024 ***Obsessions and compulsions (IG7)*** ☐

5.002	Obsessional checking	2 - 3
5.003	Obsessional orderliness	2 - 3
5.004	Obsessional thoughts of harm	2 - 3
5.005	Obsessional cleanliness	2 - 3
5.006	Obsessional ruminations	2 - 3

26.025 ***Depersonalisation and derealisation (IG8)*** ☐

3.012	Depersonalisation	2 - 3
16.006	Things appear unreal	1 - 3
16.007	People appear unreal	1 - 3
16.008	Actions unreal; acting a part	1 - 3
16.009	Self a stranger	1 - 3
16.010	Body unfamiliar	1 - 3

26.026 ***Lowered subjective functioning (IG9)*** ☐

7.002	Loss of concentration	2 - 3
7.003	Difficulty in thinking clearly	2 - 3
7.004	Loss of interest	2 - 3
7.005	Subjective retardation	2 - 3
7.006	Loss of energy	2 - 3
7.007	Loss of coping ability	2 - 3
21.030	Subjective memory problem	1 only
21.031	Compensation for 21.030	1 only
21.032	Forgets names of intimates	1 only

26.027 ***Lowered bodily functioning (IG10)*** ☐

8.005	Decreased appetite	2 - 3
8.006	Loss of weight	2 only
8.009	Poor sleep with depression	1
8.014	Early waking	2 - 3
8.025	Loss of libido	2 - 3

26.028 ***Special features of depressed mood (IG11)*** ☐

6.007	Feeling of loss of feeling	2 - 3
6.008	Unremitting depression	1
6.009	Morning depression	1
6.010	Preoccupation with catastrophe	2 - 3
6.013	Pathological guilt	2 - 3
6.014	Guilty ideas of reference	2 - 3
6.017	Loss of self-esteem	2 - 3
16.003	Dulled perception	1 - 3

26.029 ***Depressed mood (IG12)*** ☐

6.001	Depressed mood	2 - 3
6.003	Masked depression	2 only
6.004	Tearfulness	2 - 3
6.005	Loss of capacity for enjoyment	2 - 3
6.006	Loss of hope for the future	2 - 3
6.011	Suicidal ideas or actions	1 - 3
6.012	Tedium vitae	2 - 3
23.002	Behavioural depression	1 - 2

6.030 ***Depressive delusions and hallucinations (IG13)*** ☐

17.011	Type of mood congruent with AH	1 - 2
17.019	Hals associated with bereavement	1 - 3
19.010	Reference delusions: depressive	1 - 2
19.025	Delusions of guilt: depressive	1 - 3
19.026	Delusions: catastrophe/depressive	1 - 3
19.027	Hypochondriacal delusions/depressive	1 - 3

6.031 ***Delusions about the body (IG14)*** ☐

16.011	Dysmorphophobia	1 - 3
19.028	Other hypochondriacal delusions	1 - 3
19.031	Delusions concerning appearance	1 - 3
19.032	Delusions of depersonalisation	1 - 3

26.032 ***Heightened subjective functioning (IG15)*** ☐

10.010	Exaggerated self-esteem	2 - 3
10.012	Actions based on expansive mood	2 - 3
10.013	Decreased need for sleep	2 - 3
10.015	Increased libido	2 - 3

26.033 ***Rapid subjective tempo (IG16)*** ☐

10.004	Self-reported press of thought	2 - 3
10.005	Self-reported overtalkativeness	2 - 3
10.006	Self-reported distractibility	2 - 3
10.007	Self-reported overactivity	2 - 3
24.005	Flight of ideas	1 - 2

26.034 ***Expansive mood (IG17)*** ☐

10.001	Elated mood	2 - 3
23.003	Behavioural elation	1 - 2

26.035 ***Expansive delusions and hallucinations (IG18)*** ☐

17.011	Type of mood congruent with AH	2 - 3
19.011	Reference delusions: expansive	2
19.029	Delusions: grandiose ability	1 - 3
19.030	Delusions: grandiose identity	1 - 3

26.036 ***Overactivity (IG19)*** ☐

22.015	Distractibility	1 - 2
22.016	Restlessness	1 - 2
22.017	Agitation	1 - 2
22.018	Excitement	1 - 2
22.020	Destructive or violent	1 - 2
24.004	Pressure of speech	1 - 2

26.037 ***Altered Perception (IG20)*** ☐

16.002	Changing perceptions	1 - 3
16.004	Heightened perception	1 - 3
16.005	Changed perception of time	1 - 3
16.014	Other perceptual abnormality	1 - 3

26.038 ***Nonspecific AH (IG22)*** ☐

17.001	Presence of auditory hallucinations	1
17.003	Nonverbal auditory hallucinations	2 - 3
17.004	Frequency of verbal AH	2 - 4
17.005	Length of AH utterances	3 - 4
17.012	Special features of AH	1

26.039 ***Nonspecific VH (IG22)*** ☐

17.015	Unformed visual hallucinations	1 - 3
17.016	Formed visual hallucinations	1 - 3
17.017	Scenic visual hallucinations	1 - 3

26.040 ***Nonspecific psychotic (IG23)*** ☐

3.013	Non-psychotic morbid jealousy	2 - 3
16.001	Unusual perceptions	1
17.018	Hypnagogic hallucinations	2 - 3
17.022	Olfactory hallucinations	2 - 3
17.030	Hallucinations NOS.	2 - 3
19.002	Probe for other delusions	2 - 3
24.013	Magical thinking	2 - 3

26.041 ***Non-Affective AH (IG24)*** ☐

17.008	Commentary on thoughts/actions	3
17.009	Second and third person AH	3 - 5
17.010	Congruence of AH with mood	3 - 5

26.042 ***Experience of disordered form of thought (IG25)*** ☐

18.004	Loud thoughts	1 - 3
18.005	Thought echo	1 - 3
18.006	Thought insertion	1 - 3
18.007	Thought broadcast	1 - 3
18.008	Thought commentary	1 - 3
18.009	Thought block	2 - 3
18.010	Thought withdrawal	1 - 3
18.011	Other thought interference	2 - 3

26.043 ***Delusions of control (IG26)*** ☐

18.012	Replacement of will	1 - 3
18.013	Replaced control of voice	1 - 3
18.014	Replaced control of handwriting	1 - 3
18.015	Replaced control of actions	1 - 3
18.016	Replaced control of thought	1 - 3
18.017	Other replaced control	2 - 3

26.044 ***Bizarre delusions and interpretations (IG27)*** ☐

17.023	Bizarre delusions with smell	1 - 3
17.027	Delusions about sexual hallucinations	1 - 3
17.028	Hallucinations of other senses	1 - 3
17.029	Delusions with somatic hallucinations	2 - 3
18.001	Delusional perplexity	2 - 3
19.009	Delusional perception	2 - 3
19.019	Fantastic delusions	1 - 3
19.037	Incongruity of delusions	3

26.045 ***Miscellaneous delusions (IG28)*** ☐

17.025	Delusion: others think R smells	1 - 3
18.003	Delusion: thoughts being read	1 - 3
19.014	Delusional jealousy	1 - 3
19.016	Delussions of pregnancy	1 - 3
19.018	Delusion: accused of homosexuality	1 - 3
19.020	Preoccupied with past delusions	1 - 3
19.022	Delusional paranormal explanations	1
19.023	Delusional physical explanations	1
19.017	Delusional lover	1 - 3
19.021	Religions delusions	1

26.046 ***Delusions of reference (IG29)*** ☐

19.003	Delusions: being spied on	1 - 3
19.004	Delusions of reference	1 - 3
19.005	Delusional misinterpretation	1 - 3
19.006	Delusion that ideas are quoted	1 - 3
19.007	Delusional misidentification	1 - 3
19.008	Familiar people impersonated	1 - 3

26.047 ***Delusions of persecution (IG30)*** ☐

19.012	Delusions of persecution	2 - 3
19.013	Delusions of conspiracy	2 - 3

26.048 ***Emotional turmoil, etc. (IG31)*** ☐

23.004	Hostile irritability	1 - 2
23.005	Perplexity	1 - 2
23.006	Lability of emotions	1 - 2
23.007	Emotional outbursts	1 - 2

26.049 ***Incoherent speech (IG32)*** ☐

15.002	Language: speech incoherent	1 - 2
24.007	Rambling speech	1 - 2
24.011	Neologisms	1 - 2
24.012	Incoherence of speech	1 - 2

26.050 ***Other speech abnormality (IG33)*** ☐

24.001	Voice too loud	1 - 2
24.002	Voice too quiet	1 - 2
24.003	Mumbles to self	1 - 2
24.006	Circumstantial speech	1 - 2
24.008	Perseveration	1 - 2
24.009	Rumination	1 - 2
24.010	Answering past the point	1 - 2
24.014	Blocking of speech	1 - 2
24.022	Non-social speech	1 - 2
24.023	Stilted speech	1 - 2

26.051 ***Socially embarrassing behaviour (IG34)*** ☐

22.035	Odd appearance	1 - 2
22.036	Embarrassing behaviour	1 - 2
22.037	Bizarre behaviour	1 - 2
22.038	Irreverence	1 - 2
22.039	Histrionic behaviour	1 - 2

26.052 ***Flat and incongruous affect (IG35)*** ☐

23.008	Blunting of affect	1 - 2
23.009	Incongruity of affect	1 - 2
24.029	Lack of direction or purpose	2 - 3

24.053 ***Poverty of speech (IG36)*** ☐

15.005	Language: poverty of speech	1 - 2
24.015	Poverty of content of speech	1 - 2
24.016	Quantity of speech	1 - 2
24.017	Muteness	1 - 2

26.054 ***Nonverbal Communication (IG37)*** ☐

22.010	Negative posture	1 - 2
22.011	Little use of gesture	1 - 2
22.012	Little facial expression	1 - 2
22.013	Monotonous speech	1 - 2
22.014	Poor use of eye contact	1 - 2
24.018	Nonverbal communication	1 - 2
24.019	Inability to converse	1 - 2
24.020	Unrewarding response to speech	1 - 2
24.027	No 'theory of mind'	1 - 2

26.055 ***Self neglect (IG38)*** ☐

22.040	Self neglect: clothes	1 - 2
22.041	Neglect of health care	1 - 2
22.042	Neglected of hygiene and habits	1 - 2
22.045	Neglect of common dangers	1 - 2

26.056 ***Motor retardation (IG39)***

22.001	Bowed head	1 - 2
22.002	Stillness	1 - 2
22.003	Slow to move	1 - 2
22.004	Delayed start to movement	1 - 2
22.005	Underactivity	1 - 2
22.006	Stupor	1 - 2
22.007	Slow speech	1 - 2
22.008	Shuffling gait	1 - 2
22.009	Arms not swung when walking	1 - 2

26.057 ***Catatonic behaviour (IG40)***

22.021	Negativism	1 - 2
22.022	Ambitendence	1 - 2
22.023	Forced grasping	1 - 2
22.024	Echopraxia	1 - 2
22.025	Flexibilitas cerea	1 - 2
22.026	Opposition	1 - 2
22.028	Freezing in one posture	1 - 2
22.029	Automatic compliance	1 - 2
22.030	Prominence of catatonia	1 - 2

26.058 ***Monosymptomatic delusion(s) (IG41)***

19.034 rate:

1 Probably present.
2 Definitely.

26.059 ***Adequacy of information rated in IGC***

0 Good information for all significant ratings.
1 Fairly good information.
2 Much caution necessary when interpreting the ratings.
3 Very little confidence can be placed in the information available.

27 Clinical History Schedule

Checks on PSE10 and IGC ratings

Take the opportunity, while reviewing all available information on the clinical history, to check the ratings of history items in PSE10 and IGC. Consider the following in particular:

Section 0 (SCAN Face Sheet) information.

Section 1 (Episodes [also re-rated within Sections if date vary], Medication, Clinical History) dates and other clinical data.

Dates of first onset and episode dates in PSE10 Sections.

Attributions of organic and other causes in relation to PSE10 Sections in Section 13/20.

Items that relate one section to another. For example, **6.022** (anxiety and depression), which can also be reviewed in Sections 13 and 20.

Items for rating aspects of the clinical course. For example, **6.038–6.043** (episodes of depression).

Rating the Clinical History Schedule (CHS):

The PSE and IGC systematise and operationalise the collection of information relevant to disorders in sections F0 to F5 of ICD-10 and their equivalents in DSM-IIIR. They also provide opportunities for rating specific aspects of the clinical course.

The CHS is intended to supplement this information by providing an opportunity to check or enter data relevant to the broader clinical and social history and also to rate the presence or absence of disorders in ICD-10 F6–F9.

Several of the items involve rating R's performance relative to an average defined by local expectation. This involves an informed judgement, which can only be made by someone familiar with the culture and environment in which the respondent lives. See Glossary comment.

If the CHS is used, item **0.015** on the Face Sheet should show the entry – 1.

Cut-off points are included in Sections 1, 3 and 5 of the CHS. All other CHS Sections should be completed in full.

Ratings are specified in the text. With the exception of continuous variables.

8 Not known.
9 Not applicable.

1. CHILDHOOD AND EDUCATION TO AGE 16

27.001 ***Age leaving full time education***

27.002 ***Performance in school***

1 Higher than average on leaving school.
2 Achievement at level of local cultural average.
3 Lower than average but acceptable.
4 Markedly lower than average.

27.003 ***Attendance at special school***

1 None (excludes 9).
2 For maladjustment.
3 For delinquency.
4 For learning problems.
5 Other: specify.............
9 Relevant, but no specialist facilities available.

27.004 ***Psychiatric problems before 16 years old***

1 No significant psychiatric problem.
2 Information not required.
3 Significant problems rated.

Skip => to 27.024 if 27.004 is rated 1 or 2.

Rate items **27.005 – 27.018** as follows:

0 Absent or not clinically signiflcant.
1 Present but not severe.
2 Present in severe form or with severe consequences.

Behavioral and emotional disorders

27.005 ***F90*** ***Hyperkinetic, attention deficit disorder.***

27.006 ***F91*** ***Conduct disorder.***

27.007 *F92* *Mixed conduct and emotional disorder.* ☐

27.008 *F93* *Anxiety disorders.* ☐

27.009 *F94* *Elective mutism, attachment disorders.* ☐

27.010 *F95* *Tic disorders, including Tourette's syndrome.* ☐

27.011 *F98* *Enuresis, encopresis, pica, stuttering.* ☐

Specific developmental disorders

27.012 *F80* *Specific developmental disorder of articulation, expressive language, receptive language.* ☐

27.013 *F81* *Specific developmental disorder of reading, spelling, arithmetic.* ☐

27.014 *F82* *Specific motor developmental disorder.* ☐

Autistic spectrum: F84.0–1 Autism; F84.5 Asperger syndrome

Autistic disorders are developmental but they usually lead to lifelong impairments. Many of the motor, speech, affective and social behavioral anomalies observed in adults are rated in Section 22 – 25 items. However, without a good developmental history is difficult to make a diagnosis. The following items allow the four key elements of the early history to be rated. They are present from an early age and can be present at any level of intelligence.

27.015 ***Impairment of reciprocal social interaction. Poor understanding of others thoughts and feelings.*** ☐

27.016 ***Impairment of communication. Verbal language can be adequate, but content is repetitive and restricted to R's own interests, regardless of listener. Non-verbal communication is poor.*** ☐

27.017 ***Impaired development of creative imagination. History of repetitive and limited pretend play, or no symbolic play at all.*** ☐

27.018 ***These abnormalities give rise to a pattern of behavior that is dominated by repetitive and stereotyped routines, varying in complexity from simple bodily movements (e.g. hand flapping) to an exclusive interest in railway timetables, the classification of birds of prey, or drawing robots.*** ☐

27.019 ***Diagnosis of autism or Asperger syndrome previously made?***

0 No.
1 Autism (F84.0–1).
2 Asperger syndrome (F84.5).
3 Other variant of F84.

2. INTELLECTUAL LEVEL

27.020 ***Test estimate of IQ at peak***

Test:..

Date:..

27.021 ***Non-test estimate IQ at peak***

1 Under 20 (Profound).
2 20–34 (Severe).
3 35–49 (Moderate).
4 50–69 (mild).
5 70–84.
6 85–99.
7 100+.
8 NK.

27.022 ***Test estimate of IQ at key date***

Test:..

Date:..

27.023 ***Non-test estimate IQ at key date***

1–7 as **27.021** above.
8 NK.

Rate the physical cause of pathology at items **27.083–27.088**.

3. SOCIAL ROLES AND PERFORMANCE

This section deals with occupation, and social and sexual relationships. Use local standards to judge likely expectation. If there are no local opportunities and an estimate of local expectation is impossible, enter 9.

27.024 ***Social roles and performance***

1 No significant problem.
2 Information not required.
3 Significant problems rated.

Skip => to 27.049 if 27.024 rated 1 or 2.

Performance in adult occupation and education

27.025 ***Academic promise in higher education***

1 Higher than average on leaving school.
2 Achievement at level of local cultural average.
3 Lower than average but acceptable.
4 Markedly lower than average.

27.026 ***Peak paid occupation***

1 Higher than average.
2 Achievement at level of local cultural average.
3 Lower than average but acceptable.
4 Markedly lower than average.

27.027 ***Present paid occupation***

1 Higher than average.
2 Achievement at level of local cultural average.
3 Lower than average but acceptable.
4 Markedly lower than average.

27.028 ***Household duties***

1 Higher than average.
2 Achievement at level of local cultural average.
3 Lower than average but acceptable.
4 Markedly lower than average.

27.029 ***Overall occupational performance***

[Progress and stability by local standards.]

1 Higher than average.
2 Average work history.
3 Lower than average.
4 Markedly lower than average.

Adult social relationships

[Rate using local standards of judgement.]

27.030 ***Stability of social relationships at peak*** ☐

1 Higher than average.
2 Average.
3 Lower than average.
4 Markedly lower than average.

27.031 ***Stability of social relationships at present*** ☐

1 Higher than average.
2 Average.
3 Lower than average.
4 Markedly lower than average.

27.032 ***Quality of social relationships at peak*** ☐

1 Higher than average.
2 Average.
3 Lower than average.
4 Markedly lower than average.

27.033 ***Quality of social relationships at present*** ☐

1 Higher than average.
2 Average.
3 Lower than average.
4 Markedly lower than average.

27.034 ***Hypersensitivity to interpersonal rejection*** ☐

Long-standing pattern of sensitivity to interpersonal rejection not limited to episodes of mood disturbance.

0 Trait absent.
1 Trait present but little interference with social or occupational functioning.
2 Moderate or intermittently severe interference.
3 Severe interference in social or occupational functioning.

Sexual relationships

27.035 ***Stability of sexual relationships at peak***

1 Higher than average.
2 Average.
3 Lower than average.
4 Markedly lower than average.

27.036 ***Stability of sexual relationships at present***

1 Higher than average.
2 Average.
3 Lower than average.
4 Markedly lower than average.

27.037 ***Quality of sexual relationships at peak***

1 Higher than average.
2 Average.
3 Lower than average.
4 Markedly lower than average.

27.038 ***Quality of sexual relationships at present***

1 Higher than average.
2 Average.
3 Lower than average.
4 Markedly lower than average.

The subject is unable to participate in a sexual relationship as he she would wish. Dysfunction occurs frequently, has been present for 6 months or more. Rate physical, organic or psychosocial attribution in Section 13. Apply full ICD-10 DCR rules for each type of disorder. F52.10 to F52.9.

Rate items **27.039 - 27.048** as follows:

0 Absent.
1 Present but infrequent problem.
2 Frequent but may be absent on some occasions.
3 As 1 for at least 6 months.
4 As 2 for at least 6 months.

Record onset and interference with activities in Section 8.

27.039 ***F52.0*** ***Lack or loss of sexual drive.***

27.040 ***F52.1*** ***Sexual aversion and lack of sexual enjoyment.***

27.041 ***F52.2*** ***Failure of genital response.***

27.042 ***F52.3*** ***Orgasmic dysfunction.***

27.043 ***F52.4*** ***Premature ejaculation.***

27.044 ***F52.5*** ***Nonorganic vaginismus.***

27.045 ***F52.6*** ***Nonorganic dysparunia.***

27.046 ***F52.7*** ***Excessive sexual drive.***

27.047 ***F52.8*** ***Other nonorganic sexual dysfunction.***

27.048 ***F52.9*** ***Unspecified nonorganic sexual dysfunction.***

4. OVERALL SOCIAL HANDICAP (DISABLEMENT)

Check items **1.045 - 1.049:** dates of onset, first recognition, first contact and premorbid condition

27.049 ***Social performance before onset***

Level of all-round social performance before first onset/contact. Estimate taking into account average level of social and occupational performance for age, sex and local expectation.

1 High above average.
2 Good average.
3 Average.
4 Poor. Moderately disabled.
5 Severely disabled.
6 Incapacitated.

27.050 ***All-round social performance at key date***

Rate 1–6, using definition and scale of **27.039.** Include self-care and other relevant ratings from Sections 21 - 25.

27.051 ***Percentage of time socially handicapped***

What percentage of the time between first onset/contact and key date was R moderately, severely or very severely socially handicapped? Even if this estimate is approximate, a judgement will be of value.

27.052 ***Cause of social handicap***

To what extent is the degree of social handicap expressed in **27.050** and **27.051** associated with or caused by mental disorder (including mental retardation, dementia, organic and functional psychosis however caused, neurosis and personality disorder).

1 Most handicap caused by other factors, such as social disadvantage, physical disability, social effects of substance misuse, or personal attitude.
2 Factors 1 and 3 combined. Difficult to separate.
3 Mental disorder the chief factor causing social handicap.

5. DISORDERS OF ADULT PERSONALITY AND BEHAVIOR

These disorders are defined in ICD-10-DCR and the International Personality Disorder Examination (IPDE Module). The latter should be used if detailed ratings are required. The categories are listed below for convenience.

27.053 ***Personality disorders***

0 No significant problem.
1 IPDE used instead of section.
2 Information not required.
3 Significant problems rated.

Skip => to Subsection 6 ('Quality of Data Recorded in Scan,' below) if rated 1 – 3.

Personality disorders

Rate the following on the basis of personality traits beginning before the age of 18 and continuing into adult life.

0 Normal limits.
1 Conspicuous exaggeration of a single trait or lesser abnormality of several traits, but not sufficient to make a diagnosis of personality disorder.
2 Pervasive abnormality of behavior and attitudes affecting a broad range of personal and social situations. Sufficient for the main, or a second, diagnosis.

27.054 ***F60.0*** ***Paranoid Personality Disorder.***

27.055 ***F60.1*** ***Schizoid, excludes Schizotypal and Asperger disorders.***

27.056 ***F60.2*** ***Dystocial, includes sociopathic, amoral, asocial, psychopathic disorders.***

27.057 ***F60.3*** ***Emotionally unstable, includes explosive, aggressive and borderline pesonality disorder.*** ☐

27.058 ***F60.4*** ***Histrionic, includes hysterical personality disorder.*** ☐

27.059 ***F60.5*** ***Anankastic, includes obsessive/compulsive personality disorder.*** ☐

27.060 ***F60.6*** ***Anxious/avoidant personality disorder.*** ☐

27.061 ***F60.7*** ***Dependent, includes asthenic, inadequate personality disorder.*** ☐

27.062 ***F60.8*** ***Other personality disorder, specify:*** .... ☐

Enduring personality changes

27.063 ***F61/62*** ***Includes personality changes following catastrophic experiences or psychiatric illness, mixed and other changes. A diagnosis should not be made without carefully following the criteria laid down in the DCR.*** ☐

Habit and impulse disorders

27.064 ***F63*** ***Habit and impulse disorders Include pathological gambling, fare-setting, pulling out hair and stealing. A diagnosis should not be made without carefully following the criteria laid down in the DCR.*** ☐

Gender identity disorders

27.065 ***F64*** ***Includes transsexualism, dual role transvestism and gender identity disorder. A diagnosis should not be made without carefully following the criteria laid down in the DCR.*** ☐

Disorders of sexual preference or orientation

27.066 ***F65*** ***Includes: fetishism, fetishistic, transvestism, exhibitionism, voyeurism, pedophilia and sadomasochism. A diagnosis should not be made without carefully following the criteria laid down in the DCR.*** ☐

27.067 ***F66*** ***Includes sexual maturation disorder, ego dystonic sexual orientation and sexual relationship disorder. A diagnosis should not be made without carefully following the criteria laid down in the DCR.*** ☐

Other disorders of personality

27.068 ***F68.0*** ***Factitious disorder (deliberate feigning), includes Munchhausen's syndrome. Rate 1 if physical only; 2 if psychological only; 3 if both. Check item 2.052.*** [ ]

27.069 ***F68.1*** ***Elaboration of physical symptoms for psychological reasons. Check item 2.051.*** [ ]

6. QUALITY OF DATA RECORDED IN SCAN

Is the information collected of sufficient quantity and quality to allow a satisfactory diagnostic classification?

27.070 ***Quality of SCAN data*** [ ]

0 Good quality data throughout.
1 Some problems but generally acceptable.
2 Definite areas of doubt. Interpret with caution.
3 The information can only be used as a highly approximate definition of psychopathology and little weight should be placed on any diagnostic classification

7. OVERALL CLINICAL DIAGNOSES

The SCAN system provides, as one of its outputs, a classification according to the rules laid down in various nosologies. This is a 'reference classification rather than a clinical diagnosis. It is recommended that an independent clinical diagnosis be made as well in all cases. Comparison between the two classifications will often be of interest, both from a scientific and a case-management point of view. Provision is made below for three types of diagnostic system to be used.

For ICD-10, enter chapter letter, followed by up to 4 digits, starting in the left box, e.g. F22 = [F22], F227.00 = [F22 00]. For DSM-III-R, enter up to 5 digits, e.g. 295.30 = [295 30]

Diagnosis at key date

27.071 ICD-10 first diagnosis [ ][ ][ ] [ ][ ]

27.072 ICD-10 second diagnosis [ ][ ][ ] [ ][ ]

27.073 DSM-III-R first diagnosis [ ][ ][ ] [ ][ ]

27.074 DSM-III-R second diagnosis [ ][ ][ ] [ ][ ]

27.075 Local system (specify) first ....

27.076 Local system (specify) second ....

Diagnosis of previous episode/s

27.077 ICD-10 first diagnosis

27.078 ICD-10 second diagnosis

27.079 DSM-III-R first diagnosis

27.080 DSM-III-R second diagnosis

27.081 Local system (specify) first diagnosis ....

27.082 Local system (specify) second diagnosis ....

Physical illnesses or disabilities not entered elsewhere

Physical disorders thought to contribut to the causation of sections of symptoms rated in PSE10 have already been entered, for example at **2.078** for dissociative disorders, item **4.062** for anxiety disorders, etc. There is no need to enter them again.

Physical diseases occurring during the year before interview have been recorded at items **2.004** to **2.006** and the limitations placed on activities or well-being rated at item **2.008.**

If there are any other physical disorders relevant to R's health status, they should be entered below, in order of clinical significance. Enter the letter identifying the ICD-10 chapter followed up by to 4 digits. For example, conditions such as foetal rubella, Down syndrome, etc; thought to be relevant to mental retardation rated at items **27.020 – 27.023**, would be entered here, as would any physical diseases or disabilities relevant to other parts of the CHS, or not recorded elsewhere in SCAN.

Use the following rating points:

1 Present but has no influence on psychiatric status.
2 Present and has affected the social outcome but is not a cause and has not modified the clinical symptoms.
3 Present and has modified the clinical picture but is not a cause.
4 Present and is a contributory cause of the psychiatric condition.

First enter the ICD-10 code as in item **27.071**. If not known, write the name. Then enter the rating from the scale.

27.083 Condition 1 .... ☐☐☐ ☐☐

27.084 Rating .... ☐

27.085 Condition 2 .... ☐☐☐ ☐☐

27.086 Rating .... ☐

27.087 Condition 3 .... ☐☐☐ ☐☐

27.088 Rating .... ☐

The clinical course

Selected clinical episodes have been rated and dated in PSE10 and/or IGC. Information about type of onset and pre-onset social competence is recorded at items **1.048 - 1.049**. The quality of remissions between episodes is entered, if relevant, at **1.018**. The CHS allows the recording of data concerning childhood, intellectual capacity, personality, social performance and disablement, and change in disability over time. It would be useful to make a final check of these.

END OF SCAN MANUAL

Participants in SCAN field trial centers

Ankara

Hacettepe University, Faculty of Medicine, Department of Psychiatry 56, Ankara 06100, Turkey
Dr A. Gvüs, Dr TB Üstün, Dr A Usus¡ahin, Dr KR Özbayrak, Dr M Rezaki, Dr MK Yazici, Dr F Dereboy, Dr C Kiliç

Athens

Department of Psychiatry, Eginition Hospital, Athens University Medical School, 74 Vasillissis Sophias Avenue, Athens, Greece
Dr C Stefanis, Dr V Mavreas, Dr V Kontaxakis, Dr M Economy, Dr D Ploumbidis

Bangalore

National Institute of Mental Health and Neuro Sciences, P.O.Box no. 2900, Bangalore 560029, India
Dr M Isaac, Dr S Chatterjee, Dr TG Sriram, Dr C Chandrashekar, Dr RS Murthy

Beijing

Institute of Mental Health, Beijing Medical College, Beijing, People's Republic of China
Dr She Liang, Dr Xu Yu-Xin

Canberra

Social Psychiatry Research Unit, The Australian National University, A.C.T. 2601, Australia
Dr B Hughson, Dr AS Henderson, Dr M Pentony, Dr V Kingham, Dr M Christie, Dr M Hayes, Dr MSR Scott

Farmington

School of Medicine, Department of Psychiatry, University of Connecticut Health Center, Farmington, Ct 06032, USA
Dr J Escobar, Dr T Babor, Dr J Brown, Dr R Salomon, Dr N Liebowitz, Dr M Bohn

Geneva

Clinic Psychiatric Cantonal, 2018 Perreux, Switzerland
Dr L Barrelet, Dr M Touabi

Groningen

Psychiatric Clinic, Academisch Ziekenhuis Groningen, Oostersingel 59, Postbus 327.0001, 9700 Groningen, Netherlands
Dr W van den Brink, Dr RJ van den Bosch, Dr F Nienhuis, Dr R Giel, Dr C Slooff, Dr L Mulder

Leicester
Department of Psychiatry, University of Leicester, Clinical Sciences Building, Leicester Royal Infirmary, PO Box 65, Leicester LE2 7LX, UK
Dr T Brugha, Dr N Agnew, Dr J Bruce, Dr S Deb, Dr K de Pauw, Dr S Fallow, Dr C Maloney, Dr E Napier, Dr S Nayani, Dr J Warrington

London
MRC Social Psychiatry Research Unit, Institute of Psychiatry, DeCrespigny Park, London SE5 8AF, UK
Dr P Bebbington, Dr L Appleby, Dr M Perkins, Dr P Silverstone

Lübeck
Klinik für Psychiatric der Medizinischen Universität, Ratzeburger Allee 160, 2400 Lübeck, Germany
Dr H Freyberger, Dr H Dilling, Dr V Dittman

Mannheim
Central Institute of Mental Health, J.5/POB 122220, 6800 Mannheim 1, Germany
Dr K Maurer, Dr H Häfner, Dr A Hillig, Dr R Olbrich, Dr S Velthaus, Dr M van Gülick-Bailer

Nottingham
Department of Psychiatry, Queen's Medical Centre, Nottingham NG7 2UH, UK
Dr G Harrison, Dr T Friedman, Dr P Mason, Dr I Medley

Santander
Social Psychiatry Research Unit of Cantabria, Hospital Nacional 'Marques de Valdecilla', 39008 Santander, Spain
Dr JL Vázquez-Barquero, Dr JF Diaz Manrique, Dr L Gaite, Dr T Arenal, Dr S Herrera Castenedo, Dr C Peña, Dr J Artal

Sao Paolo
Departamento de Psiquiatria, Faculdade de Medicina de Universidade de Sao Paolo, Avenida Dr Arnaldo 455 – CEP 01246, Caixa Postal 8921, Sao Paolo – SP Brazil
Dr L Andrade, Dr F Lotufo, Dr V Gentil, Dr H Vallada

Sofia
Neuroscience and Behavior Research Program, WHO Collaborating Centre for Mental Health, Medical Academy, Boulevard D Nestorov 15, Sofia 1431, Bulgaria
Dr V Nikolov, Dr T Tomov, Dr K Ganev, Dr A Jablensky

Sydney
Clinical Research Unit for Anxiety Disorders, St Vincent's Hospital, 299 Forbes Street, Darlinghurst, NSW 2010, Australia
Dr G Andrews, Dr R Mattick

ATTRIBUTIONAL RATING SCALE (OPTIONAL)

Use the attributional rating scale to rate nature of influence on symptom presence or severity at the item level. The rating of 0 is the only 'negative' rating, and can be used if there is clear evidence that treatment has resulted in a reduced rating. Ratings 1 - 8 represent attributions of etiologic effect, increasing the likelihood of an item being rated or its severity being increased. These attributions must only be made for symptoms/items that have positively rated, and should be entered into the dashed boxes below the standard episode rating boxes. The rating of 9 is reserved for indicating a trait characteristic. If no effect or trait is judged to be present leave the dashed box blank.

0 Item rating reduced by effects of treatment
1 Alcohol
2 Other psychoactive substance
3 Effects of somatic psychiatric treatments (electroconvulsive therapy, antidepressant medication, neuroleptics, etc)
4 Known primary intracranial process (Alzheimer's disease, Huntington's disease, Parkinson's disease, tumor, stroke, etc.)
5 Non-psychiatric medication, toxins
6 Other medication condition 1 (specify at Section 13 and/or Section 20)
7 Other medication condition 2 (specify at Section 13 and/or Section 20)
8 Other medication condition 3 (specify at Section 13 and/or Section 20)
9 Trait. Essentially lifelong characteristic of R.

Rating Scale IV

0 Absent. Below IG threshold.
1 Subject to clinical discretion, at least two items should definitely pass PSE threshold level, PLUS at least moderate disability (interference with everyday activities) and/or moderate or severe distress resulting from the symptoms.
2 As 1, but with severe degree of distress and/or disability. Usually at least three items pass PSE threshold but this is at the interviewer's clinical discretion.
8 Uncertain whether IG is present (usually because of insufficient information).

Rating scale III Behavior, speech and affect

NB: Consider behavioral items for cognitive impairment also.

Most items in Sections 22 – 24 are rated on a 3-point scale (0–2) on the basis of severity and frequency during the past month. Information from records for the period should be used as well.

The examination should be supplemented by taking into account any other observations of relevance, e.g. is case records or information from professionals or relatives. Severe behavioral abnormalities may not be observed at examination because of the short time sample, but when they are present skilled direct observations are of great importance. The items listed are also worth rating because of their possible juxtaposition with other symptoms. The time period rated is the month before examination.

Many behavioral items are also included in Item Groups and can be rated in the Checklist.

0 Behavior not present during past month.
1 Unequivocally present during past month, moderate severity only. Use all information available.
2 Present in severe form during past month or at examination.
8 Unsure whether present or not after adequate examination.
9 Not appropriate to make a rating because examination or records are incomplete, or behavior is due to a physical factor.

Rating scale II

0 Symptom did not occur during PERIOD.
1 Symptom definitely occurred during the period but was probably uncommon or transitory.
2 Symptom was definitely present, on multiple occasions or for part of the time, during the period.
3 Symptom was present more or less continuously throughout the period.
5 Language difficulty, rated as present in Section 15, makes replies difficult to interpret.
8 Rater is unsure whether the phenomenon is present or absent, even after adequate examination
9 Not appropriate to make a rating because examination incomplete, e.g. because of refusal, omission, etc.

Rating Scale I

0 Symptom absent after adequate examination (PSE9 = 0).
1 Symptom has been present during the period but only in mild degree. Below threshold for diagnosis but counts in scores. It is not the same as 8 (PSE9 = 0).
2 Symptom definitely present but of moderately severe intensity or, if severe, present for less than half the time. If unsure whether 1 or 2, rate 1 (PSE9 = 1).
3 Severe for more than half the period. If unsure whether 2 or 3, rate 2 (PSE9 = 2).
5 Psychotic symptoms make rating of subjective symptoms difficult.
8 Not known whether the phenomenon is present or absent, even after adequate examination. This rating should not be confused with 1 (= mild).
9 Not appropriate to make a rating because examination is incomplete in some relevant respect, e.g. because of language disorder, refusal, omission, irrelevance. Examination either not possible or not necessary.

SCAN COMPUTER PROGRAM

A set of computer programs are available that allow processing of all the information obtained in use of the SCAN. Data from SCAN can also be entered in a specific computer program *(DATA ENTRY)*, or the interview can be administered using *CAPSE (COMPUTER-ASSISTED PSE)*. The output is presented as a series of options, including a range of profiles of Symptom and Item Group Scores, an Index of Definition, and ICD-10, DSM-III-R, and DSM-IV categories. In this way, a large clinical database can be generated, diagnoses can be easily checked, and complex analysis can be easily performed.

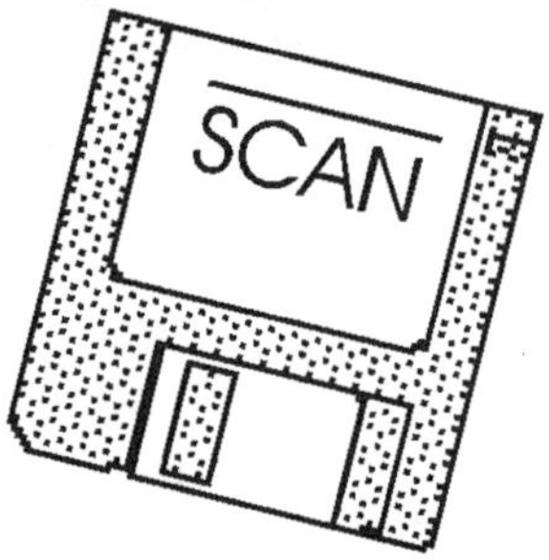

The computerization of SCAN was developed on a Windows platform, and the interview text, the Glossary, and the rating boxes are available on screen. On-line help and a tutorial are available. The ICD-10-DCR and DSM-IV criteria can be called up easily. Data can be represented graphically. The data are stored in a standard format and can be exported to a statistical analysis package. This facilitates efficient, error-free collection of data, saves time, and permits large-scale data management.

The computer program is available in English for DOS or Windows platforms (1994); versions in other languages are being prepared. To obtain the program or information about recent developments, please contact your nearest SCAN Training and Reference Center or contact

Dr. T. Bedirhan Üstün
World Health Organization
Division of Mental Health
Room L319
CH 1211 Geneva 27
Switzerland

e-mail (internet): b.ustun@who.ch
Fax: (41.22) 788.29.86